Principles of Toxicology

Anju Agarwal
Associate Professor
Department of Zoology
S.N. Sen B.V. P.G. College
CSJM University, Kanpur

Krishna Gopal
Deputy Director and Head
Aquatic Toxicology Division,
Indian Institute of Toxicology Research
(Council of Scientific & Industrial Research)
Lucknow

Published by

Khushnuma Complex, Basement,
7, Meerabai Marg,
(Behind Jawahar Bhawan),
Lucknow - 226 001, U.P., India
Tel. : +91-522-2209542/ 42/ 44/ 45
Fax : +91-522-4045308
E-mail : ibdco@airtelmail.in
Website : www.ibdco.co.in

First Edition 2010

ISBN 978-81-8189-548-6 (PB)

Composed & Designed at :
Panacea Computers
3rd Floor, Agrawal Sabha Bhawan
Subhash Mohal, Sadar Cantt. Lucknow-226 002
Phone : 0522-2483312, 9335927082, 9452295008
E-mail : prasgupt@rediffmail.com

Printed at:
Salasar Imaging Systems
C-7/5, Lawrence Road Industrial Area
Delhi - 110 035
Tel. : 011-27185653, 9810064311

प्रो. एस. वी. एस. राना
कुलपति
Prof. S.V.S Rana
Vice-Chancellor

फोन 0510 - 2320497
Phones 0510-2321390 (O)
0510-2320046 (R)
Fax 0510-2320761 (O)

बुन्देलखण्ड विश्वविद्यालय
झांसी - 284 128 (उ.प्र.) भारत
BUNDELKHAND UNIVERSITY
JHANSI - 284 128 U.P. (INDIA)

Foreword

The present text book entitled, "Principles of Toxicology", is an appreciable attempt made by Dr. Anju Agarwal, Dr. K. Gopal. There are not many books written by Indian authors on Toxicology. The books written by foreign friends are not easily understood by Indian students due to their seemingly tough language and extraneous examples. Therefore, present attempt by the author is commendable.

Toxicology, a multiphase, science, has achieved prolific progress in recent years due to its far ranging applications in biology, medicine, drug developement, biosafety and regulatory mechanisms and health risk assessment. Further, its subdivision viz., clinical toxicology, forensic toxicology, occupational toxicology, veterinary toxicology etc have now fully grown to be dealt as separate chapters. Enormous and recent interest in environmental toxicology has suppassed all dimensions of growth achieved by other branches of toxicology. Adoption of omic technologies by toxicologists further makes the subject of interest not only to human beings but for the existence of all living organisms.

This book brings topics of General and Applied Toxicology together for the convenience of the student. The figures and flowcharts with simple language make it a useful book for undergraduate, postgraduate and research students. In my opinion, the effort made by Dr. Agarwal and Dr. Gopal is praiseworthy. It should find a niche in the libraries of Indian Universities and research centres as well. I wish the reader a pleasant reading.

(Prof. S.V.S. Rana)
Vice Chancellor

Preface

Toxicology is the study of toxicants. It includes identification of poisons, their chemical properties and their biological effects. It also encompasses the treatment of disease conditions they cause. It is the science that combines biology and chemistry to study the toxicity and their effect on biological items. Toxicology is the science of adverse effects of chemicals including drugs on living organisms. It has been defined as the study of the adverse effects of Xenobiotics and is thus a borrowing science that has evolved from ancient poisons. Toxicology in this century continues to develop and expand by assimilating knowledge and techniques from most branches of biology, chemistry, mathematics and physics.

The contributions and activities of toxicologists are diverse and widespread. In the biomedical area, toxicologists are concerned with the mechanism of action and exposure to chemical agents as a cause of acute and chronic illness. Toxicologists contribute to physiology and pharmacology by using toxic agents to understand physiological phenomena. They are involved in the recognition, identification and quantification of hazards resulting from occupational exposure to chemicals and public health aspects of chemicals in air, water and other parts of the environment; foods and drugs.

In the past six decades industrialization, agricultural development and simultaneous growth in health care have changed life in various ways. Consequently average life expectancy has risen due to numerous such factors.

The whole book has been divided into 21 chapters. In the first chapter the introduction, scope and general principles of toxicology are well defined. The second deals with the methods of toxicology testing. In the third Ecotoxicology measures are taken into account. The fourth chapter takes Occupational health and safety measures into account. Occupational diseases are described in the fifth chapter. The carcinogenesis caused by environment is taken into account in the sixth chapter. The seventh chapter deals with the toxicology related to the industry. Eighth chapter incorporates the Environmental Monitoring which is divided into two parts. a) It describes chemical monitoring and b) biological monitoring. Ninth chapter deals with the Introduction to Bioinorganic Chemistry. In the tenth chapter Heavy Metal Toxicity is taken into account. Eleventh chapter illustrates cytotoxicity, genotoxicity, ecotoxicology and carcinogenesis. Environmental concepts are dealt with in the twelfth chapter. The impact of pollutant in Ecosystem is studied in the thirteenth chapter. Fourteenth chapter deals with Biodegradation, Biotransformation and Acid Rain. Solid Waste Management is defined in the fifteenth chapter. In the sixteenth chapter Agencies related to Protection of Environmental Health is taken into account. Metal lig and Interaction in Biological Fluids is dealt with in the seventeenth chapter. Eighteenth and Nineteenth chapters deal with the Environmental Impact Assessment and soil pollution. Twentieth and Twenty first chapters deal with Air,

Noise and Water pollution. Over and above two components which are toxicopharmacokinetics and toxicodynamics may ultimately address the toxicity and fate of any chemicals and also reveal a promising tool for comprehensive study.

The book is written according to the graduate and post graduate level courses of UGC and would be of great value to researchers as well. The present knowledge of Toxicology has assumed gigantic proportions and new updates arise on a day to day basis. Attempt has been made to use a simple and lucid language along with illustration and visuals in this book. We are aware of the fact that shortcomings are bound to creep in. However, we shall be grateful to our readers if they point out the lapses in the book.

Our sincere thanks to the dedicated team of IBDC Publishers for their marvelous effort in giving a shape of the book to our manuscript.

Contents

Contents

Chapter 1

Introduction : Scope and General Principles of Toxicology

The word toxicology is derived from two Greek words: toxikon meaning arrow (poison) and logos, meaning study or discourse. Toxicology is the study of adverse effects of chemicals on various biological systems including humans. In scientific terms, toxicology is the qualitative and quantitative study of adverse or toxic effect of chemicals and other arthropogenic materials or xenobiotics on organisms. Toxic effects may be expressed in several ways, including lethal effects (change in growth, development, reproduction, pharmacokinetic responses, pathology, biochemistry, physiology and behavior).

But in modern times toxicology is taken in much more wider sense than only the study of adverse effects of chemicals. Toxicology is also concerned with the concentrations or quantities of chemicals that can be expected to occur in the environment and therefore, includes also the study of the transport, distribution, transportation and ultimate fate of chemicals in the environment.

The contribution and activities of toxicologists are diverse and widespread. In the biomedical area, toxicologists are concerned with mechanism of action and exposure to chemical agents as a cause of acute and chronic illness. Toxicologists contribute to physiology and pharmacology by using toxic agents to understand physiological phenomena. They are involved in the recognition, identification and quantification of hazards resulting from occupational exposure to chemicals and the public health aspects of chemicals in air, water, other parts of environment, foods and drugs.

Toxicology like medicine is both a science and an art. The science of toxicology is defined as the observational and data gathering phase, whereas the art of toxicology consists of utilization of the data to predict outcomes of exposure in human and animal populations. In most cases, these phases are linked because the facts generated by the science of toxicology are used to develop extrapolations and hypothesis to explain the adverse effects of chemical agents in situations where there is little or no information.

History of Toxicology

Toxicology dates back to the earliest humans, who used animal venoms and plant extracts for hunting, warfare and assassination. The knowledge of these poisons must have predated recorded history. The Ebers Papyrus (Circa 1500 BC) contains information pertaining to many recognized poisons, including hemlock (the slate poisons of the Greeks), aconite (a Chinese arrow poison) opium (used as both a poison and as antidote) and metals such as lead, copper and antimony. There is also an indication that plants containing substances similar to digitalis and belladonna alkaloids were known.

Poisonings in Rome reached epidemic proportions during the fourth century BC (Livy). It was during this period that a conspiracy of women to remove men from whose death they might profit was uncovered. Similar large scale poisoning continued until Salla issued the Lex Cornelia (Circa 82 BC). This appears to be the first law against poisoning and it later became a regulatory statute directed at careless dispensers of drugs. Nero (A.D. 37-68) used poisons to do away with his step brother Brittanicus and employed his slaves as food tastes to differentiate edible mushrooms from their more poisonous kin.

Scope of Toxicology

Toxicology is the youngest science of poisons. Its scope has been enlarging. Toxicology is both a science and like medicine, an art that is practiced. The central focus of toxicology is a concern with the adverse effects of chemicals on living organisms and a commitment to assess the likelihood that such adverse effects will occur. The following definition should clarify the scope of toxicology.

Untoward effects of any substance in the environment, including drug toxicities or poisoning whether as a consequence of internal or external exposure, and relating to both immediate and long-term implications in the realm of human ecology, or to transformation which leads to an understanding of the mechanisms through which toxic manifestations arise and to the means for alleviation and control of toxic effects.

Toxicology is one of the several multidisciplinary fields of environmental science. Many of the fields are overlapping as many of the things that they derive from other disciplines are common to them.

Since toxicology also concerns with the study of concentrations, distribution, transportation and ultimate fate of chemicals in the environment, it has evolved as a multidisciplinary field of study which burrows several other basic sciences. It is necessary to understand the chemical (e.g. hydrolysis, oxidation and photolysis), physical (e.g. molecular structure, solubility, volatility and sorption) and biological (e.g. biotransformation) factors that affect environmental concentrations of chemicals in order to assess how potentially toxic agents act in the environment and how the environment acts on agents and to estimate the potential exposure of organisms. Knowledge of ecology, physiology, biochemistry, histology and behaviour is prerequisite in order to understand the effect of toxicants on organisms. Statistical analysis and mathematical modeling are used to quantify and predict biological effects and determine the probability of their occurrence. It is also essential to comprehend analytical chemistry methodologies and instrumentation to identify and quantify low concentrations of chemicals in various components of environment.

Toxicometrics is the study of measurement of poisons. Toxicodynamics deals with the biochemical and physiological effects of chemicals and mechanisms of their action. Toxicodynamics, therefore, emerged from physiology, biochemistry and pathology. Toxicokinetics or chemobiokinetics deals with the absorption, distribution and biotransformation and excretion of chemicals. Many basic principles of biochemistry and enzymology are applied to understanding of toxicodynamics.

A toxicologist is a scientist who applies and exerts knowledge to the study of adverse effects of chemicals. A toxicologist must, therefore, receive training in anatomy, histology, pathology, physiology, biochemistry, enzymology, endocrinology, genetics, microbiology, entomology and embryology. At the same time, he must also obtain skills required in pharmacology, radiation biology, analytical chemistry, instrumentation, biostatics, computer science and clinical medicine.

Principles of Toxicology

The professional activities of toxicologist fall into three main categories:

1. Descriptive
2. Mechanistic
3. Regulatory

Although each has a distinctive characteristic, each contributes to the other, and all are initially important to chemical risk assessment.

A mechanistic toxicologist is concerned with identifying and understanding the cellular, biochemical and molecular mechanisms by which chemicals exert toxic effects on living organisms. The results of mechanistic studies are very important in many areas of applied toxicology. The mechanistic data may be very useful in demonstrating that an adverse outcome (e.g. cancer, birth defects) observed in the laboratory animals are directly related to humans. For example, the relative toxic potential of organophosphate insecticides in humans, rodents and insects can be accurately predicted on the basis of understanding of common mechanisms (inhibition of acetyl cholinesterase) and differences in biotransformation for these insecticides among the different species. Similarly, the mechanistic data may be very useful in identifying adverse responses in experimental animals that may be relevant to humans.

Mechanistic data are useful in the design and production of safer alternative chemicals and in rational therapy for chemical poisoning and treatment of disease. For example, the drug thalidomide was originally marketed in Europe as a sedative agent for pregnant women. However, it was banned in 1962 because of devastating birth defects that occurred, if the drug was ingested during critical period of pregnancy. Thalidomide has been rediscovered as a valuable therapeutic agent that may be highly effective in the treatment of certain infectious diseases (e.g. Leprosy and AIDS), a variety of inflammatory diseases and some types of cancer. Its use in pregnant women is strictly prohibited but the discovery of this mechanism of action may not allow this drug to be used safely and effectively in the treatment of many other diseases. In addition to aiding directly in the identification, treatment and prevention of chemical toxicity an understanding of the mechanism of toxic action contributes to the knowledge of basic physiology, pharmacology, cell biology and biochemistry.

A regulatory toxicologist has the responsibility for deciding on the basis of data provided by descriptive and mechanistic toxicologists, whether a drug or other chemical poses a sufficiently low risk to be marketed for a stated purpose. The Food and Drug Administration (FDA) is responsible for allowing drugs, cosmetics and food additives to be sold in the market according to the Federal food, Drug and Cosmetic Act (FDCA). The US Environmental Protection Agency (EPA) is responsible for regulating many other chemicals according to Federal Insecticide, Fungicide and Rodenticide Act (FIFRA), the Toxic Substance Control Act (TSCA), the Resource Conservation and Recovery Act (RCRA), Safe Drinking Water Act and the Clean Air Act.

There are other specialized areas of toxicology such as forensic, clinical and environmental toxicology. Forensic toxicology is a hybrid of analytical chemistry and fundamental toxicological principles. It is concerned primarily with the medico legal aspects of harmful aspects of chemicals on humans and animals. The expertise of forensic toxicologists is invoked primarily to aid in establishing the cause of death and determining its circumstances in a post-mortem investigation. Clinical toxicology designates an area of professional emphasis in the realm of medical sciences that is concerned with disease caused by uniquely associated with toxic substances. Generally, clinical toxicologists are physicians who receive specialized training in emergency medicine and poison management. Efforts are directed at treating patients poisoned with drugs or chemicals and at development of new techniques to treat those intoxicants. Environmental toxicology focuses on the impacts of chemical pollutants in the environment on biological organisms. Ecotoxicology is a specialized area.

Spectrum of Toxic Dose

One could define a poison as any agent capable of producing a deleterious response in biological system, seriously injuring function or producing death. Among chemicals there is a wide spectrum of doses needed to produce deleterious effects, serious injury or death. This is demonstrated in Table-1 which shows the dosage of chemicals needed to produce death in 50 percent of treated animals (LD_{50}). Some chemicals produce death in microgram doses and are commonly thought of as being extremely poisonous. Other chemicals may be relatively harmless after doses in excess of several grams. The measures of acute lethality such as LD_{50} may not accurately reflect the full spectrum of toxicity or hazard, associated with exposure to a chemical. For example, some chemicals with low acute toxicity may have carcinogenic or teratogenic effects at doses that produce no evidence of acute toxicity.

Classification of Toxic Agents

Toxic agents are discussed in terms of their target organs (liver, kidney, haemopoietic system etc.), use (pesticide, solvent, food additive etc.), source (animals and plant toxins) and effects (cancer, mutation, liver, injury etc). The term 'toxic' generally refers to toxic substances that are produced by or by-product of anthropogenic (human-made) activities. Thus, zeralanone, produced by a mold, is a toxin, whereas dioxin [2,3,7,8-tetrachlorodibenzo-pdioxin (TCDD)] produced during combustion of certain chlorinated organic chemicals is a toxicant.

Toxic agents may be classified in terms of their physical state (gas, dust, liquid); their chemical stability or reactivity (explosive, flammable, oxidizer); general chemical structure (aromatic amine, halogenated hydrocarbon etc. or poisoning potential extremely toxic, very toxic, slightly toxic etc.). Classification of toxic agents on the basis of their biochemical mechanisms of action (e.g. alkylating agent, sulphydyl inhibitor, met hemoglobin producer) is usually more informative than classification by general terms such as irritants and corrosives. But more general classifications such as air pollutants, occupation related agents and acute and chronic poisons can provide a useful focus on specific problem. No single classification is applicable to the entire spectrum of toxic agents and that combinations of classification systems or a classification based on other factors may be needed to provide the best rating system for a special purpose. Nevertheless, classification systems that take into consideration both the chemical and the biological properties of an agent and the exposure characteristics are more likely to be useful for legislative or control purpose and for toxicology in general.

Table 1 : Approximate Acute LD_{50} of Some Representative Chemicals

AGENT	LD_{50} mg/kg*
Ethyl alcohol	10,000
Sodium chloride	4,000
Ferrous sulphate	1,500
Morphine sulphate	900
Phenobarbital sodium	150
Picrotoxin	5
Strychnine sulfate	2
Nicotine	1
d-Tuborcurarine	0.5
Hemicholinium-3	0.2
Tetrodotoxin	0.10
Dioxin (TCDD)	0.001
Bilutenium Toxin	0.00001

*LD_{50} is the dosage (mg/kg body weight) causing death in 50 percent of exposed animals.

Characteristics of Exposure

Toxic effects in a biological system are not produced by chemical agents unless that agent or its metabolic breakdown products reach appropriate sites in the body at a concentration and for a length of time sufficient to produce a toxic manifestation. Many chemicals are of relatively low toxicity in the "native" form but when acted on enzymes in the body, are converted to intermediate forms that interfere with normal cellular biochemistry and physiology. Thus a toxic response that occurs is dependent on the chemical and physical properties of the agent, the exposure situation, how the agent is metabolized by the system and the overall susceptibility of the biological system or subject. The major factor that influence toxicity for a specific chemical are the route of administration and duration and frequency of exposure.

Route and Site of Exposure

The major routes (pathways) by which toxic agents gain access to the body are the gastrointestinal tract (ingestion), lungs (inhalation), skin (topical, percutaneous or dermal) and other parenteral (other than intestinal canal routes). Toxic agents generally produce the greatest effect and the most rapid response when given directly into the blood stream (the intravenous route). The route of administration can influence the toxicity of the agents. For example, as agent that is detoxified in the liver would be expected to be less toxic when given via the portal circulation (oral) than when given in the systemic circulation (inhalation).

Occupational exposure to toxic agent most frequently results from breathing contaminated air (inhalation) and/or direct and prolonged contact of the skin with the substance (dermal exposure) whereas accidental and suicidal poisoning occurs most frequently by oral ingestion. In case of the lethal dose, it is likely that the skin provides an effective barrier to absorption of the agent. Toxic effects by any route of exposure also can be influenced by the concentration of the vehicle.

Duration and Frequency of Exposure

Toxicologists usually divide the exposure of the experimental animals to chemicals into four categories: acute, sub-actute, sub-chronic and chronic. Acute exposure is defined as exposure to a chemical for less than 24 hrs, and examples of exposure routes are intraperitoneal, intravenous and subcutaneous injection, oral incubation and dermal application. Sub-acute exposure refers to repeated exposure to chemical for 1 month or less, sub-chronic for 1 to 3 months and chronic for more than 3 months. Thus workplace or environmental exposures may be described as acute (occurring from a single incident or episode), sub-chronic (occurring repeatedly over several weeks or months), or chronic (occurring repeatedly for many months or years).

For many agents, the toxic effects that follow a single exposure are quite different from those produced by repeated exposure. For example, the primary acute toxic manifestation of benzene in Central Nervous System (CNS) leads to depression, but repeated exposures can result in bone marrow toxicity and an increased risk of leukemia. Acute exposure to agents that are rapidly absorbed is likely to produce immediate toxic effects but can also produce delayed toxicity that may or may not be similar to the toxic effects of chronic exposure.

Spectrum of Undesired Effects

The spectrum of undesired effects of the chemicals is broad. Some effects are deleterious while others are not. In therapeutics, for example each drug produces a number of effects, but usually one effect is associated with the primary objective of therapy; all the other effects are referred to as undesirable or side-effects of that drug for that therapeutic indication.

Allergic Reactions

Chemical allergy is an immunologically mediated adverse reaction to a chemical resulting from previous sensitization to the chemical or to a structurally similar one. The term 'hypersensitivity' is most often used to describe the allergic state, but allergic reaction and sensitization reaction are also used to describe this situation when pre-exposure of the chemical is required to produce the toxic effect. Once sensitization has occurred, allergic reactions may result from exposure to relatively very low doses of chemicals. However, for a given allergic individual, allergic reactions are dose-related. For example, it is well known that allergic response is an undesirable, adverse deleterious effect; it is also a toxic response. Sensitization reactions are sometimes very severe and fatal.

Idiosyncratic Reactions

Chemical idiosyncrasy refers to a genetically determined abnormal reactivity to a chemical. The response observed is usually qualitatively similar to that observed in all individuals but may take the form of extreme sensitivity to low doses or extreme insensitivity to high dose of chemicals. Patients exhibiting this idiosyncratic reaction have a genetic polymorphism in the gene for the enzyme butyryl cholinesterase, which is less active in breaking down succinyl chlorine. Family pedigree and molecular genetic analysis have demonstrated that the presence of low plasma butyryl cholinesterase activity is due to the presence of one or more polymorphisms in this gene. The genetic basis for the idiosyncratic response single nucleotide change in codon 127, which results in replacement of serine with proline, has been identified as a single nucleotide.

Immediate Versus Delayed Toxicity

Immediate toxic effects can be defined as those that occur or develop rapidly after a single administration of substance, whereas delayed toxic effects are those that occur after a lapse of some time. Carcinogenic effects of chemicals usually have a long latency period, often 20 to 30 years after the initial exposure, before tumors are observed in humans.

Reversible Versus Irreversible Toxic Effects

Some toxic effects of chemicals are reversible and others are irreversible. If a chemical produces pathological injury to a tissue, the ability of that tissue to regenerate largely determines whether the effect is reversible or irreversible. Thus, for a tissue such as liver, which has high ability to regenerate, most injuries are reversible, whereas injury to the CNS is largely irreversible because differentiated cells of the CNS cannot divide and be replaced. Carcinogenic and teratogenic effects of chemicals; once they occur are usually considered irreversible toxic effects.

Local Versus Systemic Toxicity

Local effects are those that occur at the site of first contact between the biological system and the toxicant. Such effects are produced by the ingestion of caustic substances or the inhalation of irritant materials. For example, chlorine gas reacts with lung tissue at the site of contact, causing damage and swelling of the tissue, with fatal consequences, even though very little of the chemical is absorbed in the blood stream. The alternative to local effects is systemic effects. Systemic effects require absorption and distribution of a toxicant from its entry point to a distant site, at which the deleterious effects are produced. Most substances except highly reactive materials produce systemic effects. For example, tetraethyl lead produces effects on skin at the site of absorption and then is transported systemically to produce its typical effects marked, there may be also indirect systemic effects. For example, kidney damage after a severe acid burn is an indirect systemic effect because the toxicant does not reach the kidney.

Most chemicals that produce systemic toxicity do not cause a similar degree of toxicity in all organs; instead they usually elicit their major toxicity in only one or two organs. These sites are referred to as the target organs of toxicity of a particular chemical. For example, lead is concentrated in the bone, but its toxicity is due to its effect on soft tissues, particularly the brain, DDT is concentrated in adipose tissue but produces no toxic effects in that tissue. The target organ of toxicity most frequently involved in systemic toxicity is the CNS (brain and spinal cord).

Interactions of Chemicals

As the individual comes in contact with large number of chemicals at any given time (workplace, drugs, diet, hobbies etc.), it is necessary in assessing the spectrum of responses, to consider how different chemicals may interact with each other. Chemical interactions are known to occur by a number of mechanisms, such as alterations in absorption, protein binding and the biotransformation and excretion of one or both of the interacting toxicants. The response of the organisms to combinations of toxicants may be increased or decreased because of toxicologic responses at the site of action.

The effect of two chemicals given simultaneously produces a response that may simply be additive of their individual responses or may be greater or less than that expected by addition of their individual responses. A number of terms have been used to describe pharmacologic and toxicologic interactions. An additive effect occurs when the combined effects of two chemicals is equal to the sum of the effects of each agent given alone. The effect most commonly observed when two chemicals are given together is an additive effect. For example, when two organophosphate insecticides are given together, they produce much more liver injury than the mathematical sum of their individual effects on liver at a given dose would suggest. Potentiation occurs when one substance does not have toxic effect on certain organ or system but when added to another chemical makes the chemical more toxic (example 0+2=10). Isopropanol, for example, is not hepatotoxic, but when it is administered in addition to carbon tetrachloride, the hepatotoxicity is much more greater than when it is alone. Antagonisms occur when two chemicals administered together interfere with each other's actions or one interferes with the action of the other (example 4+6=8; 4+(-4)=0; 4+0=1). Antagonistic effects of chemicals are often very desirable in toxicology and are the basis of many antidotes. There are four major types of antagonism: **functional, chemical, dispositional and receptor.**

(a) **Functional antagonism** occurs when two chemicals counterbalance each other by producing opposite effects on the same physiological function. The blood pressure can markedly fall during severe barbiturate intoxication, which can be antagonized by the intravenous administration of a vasopressor agent such as nor-epinephrine or metraminol. Similarly, many chemicals when given at toxic dose levels, produce convulsions and the convulsions can be controlled by giving anticonvulsants such as benzodizepines (e.g. diazapam).

(b) **Chemical antagonism** or inactivation is simply a chemical reaction between two compounds that produce a less toxic product. For example, dimercprol chelates with metal ions such as arsenic, mercury, lead thus decrease their toxicity. The use of antitoxins in the treatment of various animal toxins is also an example of chemical antagonism.

(c) **Dispositional antagonism** occurs when the disposition-that is, the absorption, biotransformation, distribution or excretion of a chemical is altered so that the concentration and/or duration of the chemical at the target organ are diminished. Thus the prevention or absorption of a toxicant by ipecac or charcoal and the increased excretion of a chemical by administration of an osmotic diuretic or alteration of the pH of the urine are examples of dispositional antagonism.

(d) **Receptor antagonism** occurs when two chemicals that bind to the same receptor produce less of an effect when given together than the addition of their separate effects (example; 4+6=8) or when one chemical antagonizes the effect of the second chemical (example; 0+4=1). Receptor antagonists are also called the blockers.

Tolerance

Tolerance is a state of decreased responsiveness to toxic effects of a chemical resulting from prior exposure to a chemical or to a structurally related chemical. Two major mechanisms are responsible for tolerance. One is decreased amount of toxicant reaching the site where the toxic effect is produced (dispositional tolerance) and the other is due to a reduced responsiveness of a tissue to a toxic chemical than is known about dispositional tolerance. Two chemicals known to produce dispositional tolerance are carbon tetrachloride and cadmium.

Dose Response

The characteristics of exposure and the spectrum of effects come together in a correlative relationship customarily referred to as dose-response relationship. Whenever response is selected for measurement the relationship between the degree of response of the biological system and the amount of toxicant administered assumes a form that occurs so consistently so as to be considered the most fundamental and persuasive concept in toxicology.

Individual or Graded Dose-Response Relationship

Individual dose-response relationships are characterized by a dose-related increase in the severity of response. The dose relatedness of the response often results from an alteration of a specific biochemical process.

Quantals Dose–Response Relationships

In contrast to the "graded" or continuous scale dose-response relationship that occurs in individuals, the dose-response relationships in a population are by definite quantal- or "all or none" – in nature that is at any given dose, an individual in a population is classified as "responder" or "non-responder". Although these distinctions or "quantal population" and "graded individual" dose–response relationships are useful, the two types of responses are conceptually identical.

The LD_{50} is the statistically derived single dose of a substance that can be expected to cause death in 50 percent of animals tested.

Variation in Toxic Responses

Selective Toxicity

Selective toxicology means that a chemical produces injury to one kind of organism without harming another form of life even though the two may exist in intimate contact. The living matter that is injured is termed as uneconomic form. They may be related to each other as parasite or host or may be two tissues in one organism. The biological diversity interferes with the ability of ecotoxicologists to predict the toxic effects of a chemical in one species (humans) from experiments performed in another species (laboratory animals).

Selective toxicity also can be a result of a difference in biochemistry in two types of cells. For example, bacteria do not absorb folic acid but synthesize it from p-amino benzoic acid, glutamic acid and pteridine, whereas mammals cannot synthesize folic acid but have to absorb it from diet. Thus sulfonamides, which resemble p-amino benzoic acid in both charge and dimensions, antagonize the incorporation of p-amino benzoic acid into the folic acid molecule–a reaction that humans do not carry out.

Species Differences

Although a basic concept of toxicology is that experimental results in animals when properly qualified are applicable to humans, it is important to recognize that both quantitative and qualitative differences in response to toxic substances may occur among different species. Even among phylogenetically similar species (e.g. rats, mice, guinea pigs and hamsters) large differences in response may occur. For example, the LD_{50} for the highly toxic dioxin, 2, 3, 7, 8-tetrachlorodibenzo-p-dioxin (TCDD) differs by more than 1000-fold between guinea big and hamsters.

Individual Differences in Responses

Genetic polymorphism is physiologically important and genes may be responsible for inter-individual differences in toxic responses. For example, studies in transgenic mice have shown that mice possessing one copy of mutated p53 gene are much more susceptible to some chemical carcinogens than are mice with two normal copies of the gene.

As our understanding of the human genome increases, more "susceptibility" of the genes will be discovered, and it is likely that the etiology of many chronic diseases will be shown to be related to a combination of genetics and environment. Simple blood tests may ultimately be developed that allow individual to learn whether he or she may be particularly susceptible to specific drugs or environmental pollutants.

Animal Toxicity Tests

Two main principles underlie animal toxicity testing. The **first** that effects produced by a compound in the laboratory animals, when properly qualified are applicable to humans. This premise applies to all of experimental biology and medicine. On the basis of dose per unit of body surface, toxic effects in humans are usually in the same range as those of experimental animals, probably by a factor of about 10. When one has the awareness of these quantitative differences, appropriate safety factors can be applied to calculate relatively safe doses for humans. All known chemical carcinogens in humans, with the possible exception of arsenic, are carcinogenic in some species but not in the laboratory animals. It has become evident that all chemicals carcinogenic in animals are also carcinogenic in humans is not true. However, for regulatory and risk assessment purpose, positive carcinogenicity tests in animals are usually interpreted as indicative of potential human carcinogenicity.

The **second** principle is that exposure of experimental animals to toxic agents in high doses is necessary and valid method of discovering possible hazards in humans. This principle is based on the quantal dose-response concept that the incidence of an effect in a population is greater as the dose or exposure increases. Practical considerations in the design of experimental model systems require that the number of animals used in toxicology experiments should always be small as compared with the size of human populations at risk. Obtaining statistically valid results from such small groups of animals requires the use of relatively large doses so that the effect will occur frequently enough to be detected. However, the use of high doses can create problems in interpretation if the responses obtained at high doses does not occur at low doses. Thus, for example, it has been shown that bladder tumors observed in rats fed with very high doses of saccharin will occur at lower doses of saccharin encountered in human diet.

Acute Lethality

The first toxicity test performed on a chemical is acute toxicity. The LD_{50} and other acute toxic effects are determined after one or more routes of administration in one or more species. The species most often used are the mouse and rat, but sometimes the rabbit and dog are employed.

Acute toxicity tests:-

(1) Gives a quantitative estimation of toxicity (LD_{50}) for comparison with other substances.

(2) Identify target organs and other clinical manifestations of acute toxicity.

(3) Establish the reversibility of the toxic response.

(4) Provide dose-ranging guidance for other studies.

Skin and Eye Irritations

The ability of a chemical to irritate the skin and eye after an acute exposure is usually determined in rabbits. The chemical is applied to the skin under four covered gauze patches and usually kept in contact for 4 hrs. The nature of covering patches depends on whether occlusive, semiocclusive or no occlusive tests are desired. For occlusive testing, the test material is covered with an impervious plastic sheet for semiocclusive tests; a gauze dressing may be used. The studies require that the material be applied to abraded skin. The degree of skin irritation is scored for erythema (redness), eschar (scab) and edema (swelling) formation and corrosive action. These dermal irritation observations are repeated at various intervals after the covered patch has been removed. To determine the degree of ocular irritation, the chemical is instilled into one eye (0.1 ml of liquid or 100 mg of solid) of each test rabbit. The contra lateral eye is used as the control. The eyes of the rabbits are then examined at various times after application.

Sensitization

Information about the potential of a chemical to sensitize skin is needed in addition to irritation testing for all materials that may repeatedly come in contact with the skin. Numerous procedures have been developed to determine the potential of substances to induce a sensitization reaction in humans.

Subacute (Repeated–Dose Study)

Subacute toxicity are performed to obtain information on the toxicity of chemical after repeated administration and an aid to establish doses for subchronic studies. A typical protocol is to give three or four different dosages of chemicals to animals by mixing it in their feed.

Subchronic

Subchronic exposure can last for different periods of time, but 90 days is the most common test duration. The principal goals of the subchronic study are to establish a "No Observed Adverse Effect Level" (NOAEL) and to further identify and characterize the specific organs or organ affected by the test compound after repeated administration. One may also obtain a "Lowest Observed Adverse Effect Level (LOAEL) as well as NOAEL for the species tested. The numbers obtained for NOAEL and LOAEL will depend on how closely the dosages are spared and the number of animals examined. Determination of NOAELs and LOAELs have numerous regulatory implications. For example, the EPA utilizes the NOAEL to calculate the reference dose (R_fD), which may be used to establish regulatory values for acceptable pollutant levels. An alternative to the NOAEL approach referred to as benchmark dose uses all the experimental data to fit one or more dose-response curves. These curves are used as benchmark dose that is defined as "the statistical lower bound on a dose corresponding to a specified level of risk". Although subchronic studies are frequently the primary or sole source of experimental data to determine both the NOAEL and the benchmark dose, these concepts can be applied to other types of toxicity testing protocols, such as that for chronic toxicity or developmental toxicity.

A subchronic study is usually conducted in two species (rat and dog) by route of intended exposure (usually oral). Atleast three doses are employed (a high dose that produces toxicity but does not cause more than 10 percent fatalities), a low dose that produces no apparent toxic effects and an intermediate dose, with 10 to 20 rats and 4 to 6 dogs of each sex per dose. Each animal should be uniquely identified with permanent markings such as ear tags, tattoos, or electronically corded microchip implants. Only healthy animals should be used, and each animal should be housed individually in an adequately controlled environment. Animals should be observed once or twice daily for signs of toxicity, including changes in body weight, diet consumption, changes in fur color or texture, respiratory or cardiovascular distress, motor and behavioral abnormalities and palpable masses. All premature deaths should be recorded and nercopsied as soon as possible.

For chemicals that are to be registered as drugs, acute and subchronic studies must be completed before the company can file an Investigational New Drug (IND) application with FDA. If the application is approved, clinical trials can commence. At the same time phase I, phase II and phase III clinical trials are performed, chronic exposure of the animals to the test compound can be carried out in laboratory animals, along with additional specialized tests.

Chronic

Long-term exposure studies are performed similarly to subchronic studies except the period of exposure is longer than 3 months. In rodents, chronic exposure is usually for 6 months to 2 years. Chronic studies in non-rodent species are usually for 1 year but may be longer. The length of exposure is somewhat dependent on the intended period of exposure in humans. If the agent is a drug planned to be used for short periods, such as antimicrobial agent, a chronic exposure of 6 months is sufficient, whereas if the agent is a food additive with the potential for life time, exposure in humans, a chronic study of 2 years in duration is likely to be required.

Chronic toxicity tests are performed to assess the cumulative toxicity of chemicals. Dose selection is critical in these studies to ensure that premature mortality from chronic toxicity does not limit the number of animals that survive to a normal life expectancy. Most regulatory guidelines require that the highest dose administered be the estimated maximum tolerance dose (MTD). The MTD has been defined by some regulatory agencies as the dose that suppresses body weight gain slightly (i.e. 10 percent) in a 90 day subchronic study.

Developmental and Reproductive Toxicity

The effect of chemicals on reproduction and development also needs to be determined. Developmental toxicity is the adverse effects on the developing organisms occurring anytime during the life span of the organism and may result from exposure to chemical or physical agents before conception (either parent), during prenatal development, or postnatally until the time of puberty. Teratology is the study of defects induced during development between conception and birth. Reproductive toxicology is the study of adverse effects on male or female reproductive system that may result from exposure to chemical or physical agent.

Mutagenecity

Mutagenesis is the ability to cause change in the genetic materials in the nucleus of cells in ways that allow the changes to be transmitted during cell division. Mutations can occur in either of the two types of cell types, with substantially different consequences. Germinal mutations damage DNA in sperm and ova, which can undergo meiotic division and therefore have the potential for transmission of the mutations to future generations. If mutations are present at the time of fertilization in the egg or the sperm, the resulting combination of genetic material may not be viable, and death may occur in the early stage of embryonic cell division. Alternatively, the mutation in the genetic material may not affect early embryogenesis but may result in the death of the foetus at a later developmental period, resulting in abortion. Somatic mutations refer to mutations in all other cell types and are not heritable but may result in cell death or transmission of a genetic defect to other cells in the same tissue through meiotic division.

Chapter 2

Methods of Toxicology Testing

A particular test should be widely accepted by scientific community, and be able to predict the effects of a wide range of chemicals on different organisms. Toxicity testing methods have been developed from time to time using different procedures, accordingly designated as:-

(i) Ecotoxonomic methods (single species, multi-species or community and ecosystem studies).

(ii) Acute toxicity tests (mostly producing lethal effects).

(iii) Chronic and subchronic toxicity tests (producing lethal or sublethal effects and include entire reproductive cycle, early stages of life cycle, bioaccumulation tests; biochemical and physiological observations; behavioral responses; histopathological examinations).

(iv) Special methods/tests.

Various methods of animal toxicological tests are as follows:-

1. Acute toxicity (single)
   - (a) LD50 determination (1-2 week observations)
     - (i) Two species (one non-rodent)
     - (ii) Two routes administration
   - (b) Irritation studies
     - (i) Dermal (rabbit)
     - (ii) Eye irritation (rabbit)
2. Subacute toxicity
   - (a) Duration 90 days
   - (b) Two species (usually rat and dog)
   - (c) Three dose levels
   - (d) Route of administration to intended route of exposure
3. Chronic toxicity
   - (a) Duration two years
   - (b) Species (preferably two, one non-rodent)
   - (c) Three dose level
   - (d) Route of administration to intended route of exposure

4. Special tests
   (a) Metabolism
   (b) Neurotoxicity
   (c) Reproduction teratogenicity (atleast one species)
   (d) Carcinogenicity
   (e) Mutagenecity.

Acute Toxicity Tests

They have been conducted with industrial effluents for more than 40 years, and more recently with dredged materials, drilling muds and mine tailings. These tests are determined to determine the dose or concentration of a particular test chemical or effluent or the level of an agent that will produce a specific response/ effect on a group of test, organisms during short-term exposure under controlled conditions. Since death is easily detected response, the most common acute toxicity test is the acute lethality test. The other two common tests in this category are irritation studies and potentiating or sensitizing agents.

(i) Lethality (LC50 or LD50 Tests)

This is one of the most common ways to express acute toxicity. LC50 is statistically estimated dose (concentration) necessary to kill 50% of a large population of tests species under stated conditions. Experimentally, this is achieved by administrating a chemical at graded doses to a group of organisms and then observing the resultant mortalities in a set of time period of 96 hours or so. Among the animals rats, mice, rabbits, guinea pigs and hamsters are the test species. The measure of acute toxicity most frequently used with fish and macroinvertebrates is the 96h median lethal concentration (96-h LC50). However, because death is not easily determined for some invertebrates, an EC50 (median effective concentration) is estimated rather than LC50. The effects used for estimating the EC50 of crabs, cray fish and shrimp, are immobilization and loss of equilibrium, defined as inability to maintain normal pressure. Among plants algae and duckweed (Lemna spp.) have been used as test organism.

Acute toxicity test methods may be categorized according to length of exposure, test situation, criteria of effects and test organisms. The data obtained in these tests are used to determine the NOEC (no observed effect concentration or no effect concentration), which is the maximum concentration of the test material that produces no statistically significant harmful effect on test organisms as compared to control in specific test and LOEC (lowest observed effect concentration) or MTC (minimum threshold concentration) which is the lowest concentration that has statistically significant deleterious effect on test organisms compared to control in a specific test.

(ii) Skin and Eye Tests

These tests are conducted with some chemicals that act as primary irritants. Chemicals when applied to skin or mucous membranes lead to tissue alterations of various types. When applied to the eye, the tissue response may be severe and may lead to other toxic manifestations such as corneal damage and blindness. If inhaled, it may destroy tissue in the respiratory tract. These tests are conducted in rabbits, guinea pigs and the tissue or mucous membrane damage is recorded subjectively.

(iii) Potentiation

The term **potentiation** or **sensitization** means that when two chemicals are acting on the same or different organs or systems, one chemical is made more effective in the presence of other chemical. The tests are generally conducted on guinea pigs where the animal is first introduced to the chemical and then challenged by additional doses of the compound. Tissue response develops at the site of injection or elsewhere. In man, patch testing is often used, but only after preliminary animal sensitizing tests have been conducted. Some chemicals eliciting a sensitizing test in animals could also occasionally show allergic response in man.

Subacute Toxicity Tests

These are adverse tests produced after subacute exposure, eliciting subacute toxicity. Subacute toxicity results from repeated exposure over a period of several days or months. Such studies are carried out at three or more dose levels in two or more species, as in animals, one species is non-rodent. One of the major objectives of subacute toxicity is to establish a dose which could be classified as no-effect dose or that dose which, if exceeded can be considered harmful to man. For radiation or carcinogenic agents, a no-dose effect perhaps is not relevant, since the smallest dose will have harmful effect. With very low doses of radiation or chemical carcinogens, the biological damage is not evident until fairly a prolonged exposure has occurred.

Chronic Toxicity Tests

These are conducted to determine if prolonged (i.e. chronic) exposure (usually weeks, months or years) to a chemical (at sublethal concentrations) would have significant adverse effects on organisms. Chronic exposures involve entire reproductive cycle of organisms. In this long-term toxicity, studies are performed due to subchronic exposures involving less than a complete life cycle stage of organisms. Chronic toxicity threshold concentrations for a number of selected species of ecosystem, receiving the toxic chemical is tested. Three categories of tests are commonly used to predict the chronic and subchronic effects of toxic chemicals on organisms: (i) life cycle toxicity tests (ii) sensitivity early life stages test and (iii) functional tests.

(I) **Full life cycle tests-** These tests measure the effects of chronic exposure to a chemical on reproduction, growth, survival and other parameters over one or more generations of a population of test organism. Chronic or full life cycle toxicity test is an important tool for understanding and evaluating the potential hazard of toxicants to organisms. A chronic toxicity test indicates the concentrations of chemical that will interfere with normal growth, development and attainment of reproductive potential of an organism. Generally, concentrations that produce chronic effects are lower than those producing more observable acute effects such as mortality. Therefore, chronic toxicity tests could provide a more sensitive measure of chemical toxicity than acute toxicity tests. These tests therefore permit evaluation of possible adverse effects of the chemical under conditions of long-term exposure at sublethal concentrations.

(i) **Maximum acceptable toxicant concentration (MATC)-** The threshold concentration that produces statistically significant deleterious effect is commonly expressed as MATC.

(ii) **Partial life cycle toxicity tests-** Each of the several groups of individuals of one species is exposed to a different concentration of toxicant throughout part of a life cycle which includes life stages observed to be especially sensitive to chemical exposure. Such tests require more than one year. Test can be completed in less than 15 months and all major life stages are still exposed to the toxicant. In fish, exposure begins with immature juveniles atleast two months before active gonad development, continues through maturation and reproduction and ends not less than 30 days after hatching of next generation.

(iii) **Most sensitive early life stage toxicity tests-** The early stages of fish species are exposed for 28-32 days from ova fertilization through embryonic, larval and early juvenile development. Major toxic effects measured are on survival and growth.

(iv) **Functional tests**-These tests measure the chronic toxicity of chemicals through various physiological functions or functional response of organisms. Functional responses includes physiological and behavioral responses as affected due to exposure of sub-lethal concentration of toxic chemicals. Among the types of functional responses that have been measured in different organisms exposed to toxicants are changes in blood chemistry, histology, swimming performance, avoidance, respiration, enzyme activities, sensory perception and disease resistance. Functional tests are being used in studies on bioaccumulation and biotransformation processes also.

1. **Behavioural responses:** Behaviour is the whole complex of observable, recordable or measurable activities of a living animal. Animal behavioral patterns (responses) are highly adaptive to environmental variables of physical, chemical

or biological nature. Behavioral toxicity occurs when the introduction of a chemical induces a behavioral change that exceeds the normal range of variability.

2. **Enzyme activities:** Some enzyme activities can be used as criteria of measuring toxicity of a chemical. These functional responses have been used mainly in studies on bioaccumulation, biotransformation and biomonitoring of chemical toxicants.

Lysosomal stability. Lysosomes are organelles responsible for the breakdown of endogenous and exogenous substrates in animal cells. A good regulation system of the permeability of the lysosome membrane is needed to prevent catalytic enzymes (hydrolases) from entering the cytoplasm. Under conditions of low chemical stress only small molecules can penetrate the membrane into the lysosome, under serious stress larger molecules can pass the membrane.

Mixed function oxidase (MFO) activity (Microsomal oxidations). The MFO system is a membrane bound non-phosphorylating electron transport complex that normally forms part of endoplasmic reticulum. The system metabolizes natural substrates (steroids, fatty acids) as well as xenobiotics, including pesticides. One of the MFO component is the iron–containing cytochrome P_{450}.

Metallothioneine induction. Metallothioneines are specific proteins without any enzyme functions, in which cysteine is an important constituent. Metal ions are easily bound by cystein, counteracting toxic effect.

3. **Histo-pathological examinations:** Examination of tissues from organisms after death may serve to identify the cause of death and possibly the causative agent. Most histological and histopathological studies are made on mammals. There are routine diagnostic tests to evaluate abnormal tissue changes resulting from exposure of an organism to the toxicant.

Other toxicity tests: There are several other toxicity tests, used in special studies on evaluating the effect of toxic chemical on organisms, by taking suitable parameters. These are as follows:-

1. Reproductive toxicity and teratogenecity

Reproductive toxicity means interference by an agent with the reproductive processes in one or more ways. The toxicant may lead to malformation in children, reduce chance of conception may be embryological, cause still berth, or death of the progeny. Teratology is the study of effects of intensive factors related to permanent structural or functional abnormalities arising during the period of embryo development. The reproductive toxicity experiments may provide some indication on the teratogenic potential of test chemical.

2. Carcinogenicity

Many drugs and chemicals are known for slowly developing, cumulative and irreversible organ damage. Carcinogenic effects are the most common of these effects. Chemicals are taken orally, parenterally or by inhalation. Chemicals may also be applied to the skin. The drugs or chemicals are tested *in vitro* and *in vivo*.

3. Mutagenicity

Animal experiments or bacterial test systems are used to identify strong mutagens. Several *in vitro* and *in vivo* test systems have been evaluated, but with low doses, the mutagenic effects are often missed. Due to development and use of *in vitro* bacterial bioassays much progress has been made in this area.

Bioassay in Toxicity Testing

Bioassay is frequently used in toxicity tests, chiefly aquatic toxicity. A bioassay is performed to determine the strength of the chemical from the degree of response elicited in the test organisms and not to estimate the concentration of the chemical that is toxic to those organisms. Bioassays are frequently used in the pharmaceuticals industry to evaluate the potency of vitamins and other pharmacologically active compounds. Bioassay is thus a test in which organisms are used to detect the presence or the effects of physico-chemical factors in the environment. The test involves exposure of animal to the toxicants for a definite period in the static laboratory environment and observing survival or other symptoms during a period of 96 hrs or so (for fish). Application factor can be determined for one toxicant and may be extrapolated to other animals.

Good Laboratory Practices (GLP) - GLP is concerned with the organizational process and the conditions under which laboratory studies are planned, performed, monitored, recorded and reported. GLP is not concerned with the scientific design of study. The presence of such a design, in the form of a study plan or protocol, is a prerequisite for the performance of a study under GLP. The GLP requirements deal with all related aspects of testing viz. personal, facilities, equipment, laboratory operations, test chemicals protocols, reports and disqualification of testing facilities.

Translocations of Chemicals

Most chemicals are distributed throughout the body in the water phase of the blood plasma. In order to act, therefore, unless it acts topically (locally) at the site of exposure (i.e. local effect), a chemical must first enter the blood stream for circulation systemically in the body. Once in the blood stream, it then reaches the tissues of each organ at rate determined by the blood flow through that organ and the rapidity of its passage across the capillary network into the cells of particular organ. Effective concentration of the xenobiotic depends upon the rate of accumulation in tissues, rate of metabolism (biotransformation) by the organism as well as the rate of excretion of the xenobiotic. The major routes by which xenobiotic enter the body are skin, gastrointestinal tract and lungs. Chemicals thus absorbed are eliminated from the circulation by metabolism and excretion.

Membrane Permeability and Mechanisms of Chemical Transfer

The cell membrane or plasma membrane is approximately 100A° thick, with pores of size ranging from 40A°. The lipid layer of plasma membrane is permeable to and readily permeated by lipid soluble substances. The membranes form barriers through which a chemical must pass before achieving a sufficient concentration at the cell where it produces its characteristic action. The membranes form barriers with specific permeability properties which together with the mediated transport system control the rate and extent of transfer of substances into and out of the cells and cell organelle. But tonic agents may in different ways modify the permeability of a membrane. For instance size, shape, degree of ionization and lipid solubility of ionized and non-ionized forms of a chemical are their important physico-chemical property that influences its passage across the membrane.

Mechanisms are either a passive transfer, in which the membrane is inert or specialized transfer where the membrane has active part in the transfer of chemicals across it. There may also be phagocytosis. Both non-polar and polar substances that possess sufficient lipid solubility may diffuse by a passive transfer. In active transport, substances are actively transported across the cell membrane and are presumed to pass into the cells that form a complex with the substance to be transported. The complex is presumed to be formed on one side of the membrane and they diffuse to the other side where the substance is released after which the carrier returns to the original surface to repeat the process. Several ions (Na^+, K^+, I^-) amino acids, some strong organic acids and bases and ionized forms of weak electrolyte are transported across the renal tubule, choroid plexuses and liver cells in this manner.

Membranous Barriers:- The complex membranous system of the organism's body is effectively involved in the distribution of chemicals. There exist several types of such barriers that affect the distribution of toxic agents. Biological complexity especially depend on the existence of permeability barriers across which concentrated gradients can be established and maintained.

In permeability barriers, the effective tight function occurs at the level of capillary endothelium e.g. brain, placenta and thymus barriers. These barriers restrict the exchange of certain molecules between the blood inside capillaries and the tissue fluid outside. These barriers are called as blood organ barriers.

(i) Blood brain barrier

The blood brain barrier is treated between the plasma and extracellular space of the brain either at the capillary endothelial cells or at that glial field that surrounds the brain capillaries. These barriers are not absolute barriers to the passage of chemical into the CNS but rather represent a site that is less permeable than most of the other areas of the body. If a chemical is ionized it will not enter the brain, whereas non ionized ones will enter the brain readily at a rate proportion to its lipid water coefficient. The route exit of chemicals from the cerebrospinal fluid differs from the route of entrance since the chemicals leave the cerebrospinal fluid by bulk fluid flow through the arachnoid villi.

(ii) Placental barrier

Many xenobiotics may cross the placenta and exert their toxic effect on the foetus. They may induce foetal deaths and even congenital abnormalities. The passage of chemicals across the placental membranes occur primarily by simple diffusion. However, the concentration of chemical in various tissues of the foetus will be determined by the ability of the individual to concentrate the chemicals which may be dependent upon plasma-protein concentration and binding affinity for various chemicals.

(iii) Blood test barrier

Like other barriers, blood tests barrier is a specific consequence of intercellular tight functions imposed either directly or indirectly between blood and other distinctive fluid compartment. It occurs between sertoli cells in the seminiferous epithelium. This barrier has an important protective role in safeguarding the germ line from noxious influences originating both from within and outside. The barriers also help in shielding meiotic cells from environmental mutagens and in prevention of autoimmune orchitis.

(iv) Blood urine barrier

This barrier is infact involved in general renal exertion of undesirable xenobiotics and their metabolites from the body. Toxic compound can be excreted from the plasma by passive diffusion through the renal tubules into the urine. Like other biological membranes, the tubular epithelium particularly behaves as a lipoprotein barrier allowing the transfer of lipid-soluble non-ionized molecules. Lipid soluble compounds present in the glomerular filtrate in the non-ionized form are therefore reabsorbed into the blood stream whereas the compounds of low lipid solubility are only partially reabsorbed.

(v) Blood–bile barrier

Like renal excretion the liver is the second important route of excretion of toxic chemicals and other wastes from the body. Besides working as a route of general excretion, liver also provides an effective barrier to the passage of toxic chemicals. The blood from gastrointestinal tract passes through the liver before reaching the general systemic circulation. Liver removes the compounds from the blood and prevents their distribution into the body. Compounds may be excreted directly into the bile without re-entering the blood stream to be excreted by liver cells into bile and thus passes into the small intestine. If the properties of the chemical or their metabolites are favorable for reabsorption a cycle may result, in which biliary secretion and intestinal reabsorption continue until renal excretion finally eliminates the compounds from the body.

Storage Depots

Several tissues act as storage depots in which chemicals tend to accumulate due to certain affinity for these tissues.

(i) **Body fat:-** Tissues and areas in the body such as fat tend to store some chemicals that are highly soluble in lipids. These chemicals tend to localize in the adipose tissue by partitioning between intracellular lipid and body water. For example organochlorinated pesticides such as DDT may attain an almost constant level in the body due to equilibrium between rates of absorption, biotransformation and excretion. Thus, a compound that has a high lipid/water partition coefficient may be stored in the body fat to a large extent and may not affect obese persons as much as lean ones.

(ii) **Brain tissues:-** Some xenobiotics have high affinity for the brain tissue. For example, CNS drugs like barbiturates, thiopentone and psychotropic drugs, chlordiazepoxide and chlorpromazine have been found to enter the brain rapidly after intravenous administration in animals.

(iii) **Erythrocytes:-** Pharmacological agents, like p-nitroalamine become localized in the erythrocytes by binding with haemoglobin (Hb). Similarly another toxic agent like triethyl tin has been reported to bind rapidly to rat Hb. Arsenic has been shown to bind mostly with blood Hb. Lead, cadmium ions also localize with erythrocytes.

Other storage depots: The components of some tissues have a particular affinity for certain xenobiotics. The antibiotic tetracycline taken up in bone binds with calcium. Cadmium has been found in kidney and liver. Intravenously administered charcoal, localizes in the spleen. Cobalt has high affinity for Hb, parquets accumulates in lung. Other transcellular fluids such as aqueous humor, endolymph, joint fluids do not generally accumulate chemicals and constitute only major storage depots for iodine in the body can concentrate drugs like percolate and some other ions.

Biotransformation sites:- The term biotransformation is used to include biologically catalyzed conversions of chemicals, other than normal body constituents i.e. xenobiotics, into other chemicals. Biotransformation reactions involving xenobiotics compounds and drugs are often catalyzed by enzymes that occur in the soluble mitochondrial or microsomol fractions of the liver. These enzyme systems are also present in the intestine, lung and kidney of many living species. In general biotransformation reactions are catalyzed by enzymes with a relatively low degree of substrate specificity compared to reactions catalyzed by enzyme involved in metabolism of constitutive chemicals. For example, the enzyme glucose 6 phosphatases recognizes glucose 6 phosphate and will remove phosphate group only from this compound, it will not be effective in removing the phosphate group from fructose 6 phosphate. In contrast, the less specific enzyme involved in biotransformation of xenobiotics tend to catalyze reactions after recognition of fundamental groups within a molecule, rather than recognition of entire molecule itself. Enzymatic biotransformation reactions may be broadly divided into two types:

Phase I – Non synthetic reactions involving oxidation, reduction and hydrolysis.

Phase II – Synthetic reactions involving conjugation.

Phase I reaction can occur enzymatically and non-enzymatically. These reactions convert compounds to products that undergo Phase II reactions. Phase II (conjugation) involves the combustion of phase I products which are endogenous polar or ionic.

Mixed function oxygenases (MFOs):

These microsomol enzymes also called monooxygenase are located in the smooth endoplasmic reticulum of cells. These subcellular fractions represent the pellet sedimented by centrifugation at 100,000 to 250,000 x g for 60 to 20 min. of the supernatant fluid after cell derbis, nuclei and mitochondrial and lysosomal fractions have been separated by centrifugation at 10,000xg for 20 min. The distinctive feature of MFOs is their dependence on reduced co-factor, nicotinamide adenine dinucleotide phosphate (NADPH) and the requirement for molecular oxygen.

Vertebrate liver, malpighian tubules and the digestive tracts of insects are such source of MFOs. More enzyme activity is noticed in the liver of mammals than in birds and fish. The rat liver of male seems to have more activity than that of female. The microsomol MFOs exhibit poor substrate specificity and require only high lipophilicity in the molecule to be biotransformed.

MFOs or monooxygenase are comprised of a group of hemoproteins collectively termed cytochrome P-450. The P-450 terminology is derived from the observation that these hemoprotein, in reduced form, can combine with carbon monoxide to form products that have an absorption peak at 450 nm. The activity of MFOs can be increased or decreased by specific drugs and xenobiotics.

Selective Toxicity: Selective toxicity results from biological diversity and variability in the response of cells and tissues to chemicals. This diversity may present the toxicological form predicting the effects of a chemical in one species from results of experiments performed in another species. Some chemicals are so selective that they adversely affect only one type of cell or tissue without harming others that are in close contact. For instance, a particular chemical or drug will affect only liver and not another organ in the body will affect only mitochondrion in the cell and not other component of the cell. In other words, a chemical acting on mitochondria of liver of one organism (say bird) may not be active on mitochondria on liver of another species (say man). Thus if the affected/non-affected organs, cells, tissues are in different organism, the chemical is species-specific. Selective toxicity of a chemical may be due to the following two possible mechanisms.

1. **Receptor Sites:** Selective toxicity may be due to the presence or absence of specific receptor or target sites in the exposed cell system, as selectivity indicates that the chemical reacts only with specific normal components of target of cell. The target may be vital to the cell and may be altered by the chemical so that it can no longer carry out its function. Thus its viability is altered, or the target may be protein or lipid that is not vital to function of the cell, so that the reaction between the chemical and the target does not produce a direct alteration in cell function. For instance organophosphorous pesticides like prorate, parathion etc. that cause delayed neurotoxicity in bird are themselves specific to protein fractions in the brain and spinal cord of animals as birds, in which they selectively inhibit esterase while the other enzymes remain unaffected.

2. **Translocation Factors:-** This includes factors that are responsible for the distribution and alteration of the concentration of the chemical at specific cells or tissues as selective absorption, translocation, biotransformation and excretion.

Biotransformation and its application:- Biotransformation may be defined as the biologically catalyzed conversion of one chemical into another. It should be differentiated from (1) purely physical chemical processes, and (2) metabolism. The former can also affect chemical conversions, such as photolysis and oxidation reductions but these do not involve the biocatalyst known as enzymes. The latter i.e. metabolism, though involves chemical conversion in a biological system is carried out under the influence of enzymes, the term is more appropriately used in connection with the biochemical reactions of carbohydrates, proteins, fats and other normal body constituents. The term biotransformation is used to include biologically catalyzed conversions of chemicals, other than the normal body constituents i.e. xenobiotics into other chemicals.

It has been known for many years that chemical compound can alter normal physiological functions in mammals, but in the recent past it has been recognized that contact with biological system can significantly affect the structure of chemical compound.

Anti-dotal Procedure in Toxicology

Toxic effects of a drug or chemical may be neutralized by administration of other chemical substances that work as antidote in the system. Antidotes to these toxic agents (poisons) counteract their principal effects, prevent their access to the site of action, enhance their excretion and can prevent tissue distribution and death. The antidotes are:

(i) Antidotes which form complexes with poison: Poisoning by metals, pesticides, bacterial toxins, drugs and other chemicals is treated by use of some antidotes that form complexes with them. For instance in case of metals, antidotes react with them to form tightly bound complexes. Thus they act by preventing or reversing the binding of toxic metals to body legends. The antidotes form with metals a specific complex called chelate. A chelate is a complex formed, between a metal and compound containing two or more potential (e.g. Ethylenediamine) . The product of such a reaction is a heterocyclic ring metal poisoning by arsenic, mercury, lead, iron, copper, thallium etc. is treated by this principle. Some pesticides, bacterial toxins and drugs are also made ineffective by antidote therapy utilizing this mechanism.

(ii) Blocking of metabolic conversion of less toxic precursor to potential poison: The treatment of methanol, selenoxystathionine and fluroacetate is done by this method. Methyl alcohol poisoning is principally due to consumption of alcohol beverages, which have been adulterated with methanol. Besides this, methanol is widely used solvent in paints, varnishes and shellacs. The adverse effects of methanol are due to its metabolites i.e. formaldehyde and formic acid. Formaldehyde selectivity damages retinal cells causing blindness, formic acid produces acidosis. Ethanol is administered as an antidote. Ethanol inhibits the metabolic oxidation of methanol. Ethanol and methanol are oxidized by the same system. Thus, ethanol results in competitive substrate inhibition to block the conversion of methanol to more toxic formaldehyde and formic acid.

(iii) Acceleration of poison excretion: The antidotal principle is used in treatment of bromide and some radionucleotide poisoning. Some chemicals accelerate the excretion of poisons. For instance, removal of bromide ions from the body can be accelerated by administration of chloride ions. Increased chloride intake results in chloride diuresis. The excretion of bromide will be approximately in the same ratio to that of chloride as present in the extracellular fluid.

(iv) Competitive action on receptors: Poisoning from neuromuscular blocking agents is the result of their injudicious clinical use. The neuromuscular blockage provides the basis for their antidotal action. For example, D-tubocurarine, the active principle of curare combines with the cholinoceptive sites at the postjunctional membranes and thereby blocks the action of acetylcholamine. Thus in presence of curare, the unusual amounts of acetylcholamine released by motor nerve terminal no longer produce a local depolarization at the end plate to generate a muscle action potential. The most useful drugs against the poisoning of such chemicals are anti-cholinesterase agents like neostigmine and physostigmine.

The action of morphine and other opium alkaloid drugs have been extensively studied. Several drugs compete for receptor site to antagonize the analgesic and narcotic effects of morphine and related drugs. For instance, antidote action of narcotic antagonists like naloxane can be life saving by morphine. It reverses all narcotic effects. This is a safe drug and occasionally may cause slight drowsiness.

(v) Blocking of receptors: The toxic effect of several organophosphorous pesticides as malathion, through cholinesterase inhibitors may be prevented by administration of potent antidotes. A good example of effective antidote is atropine. It occupies acetylcholine receptors, rendering them inaccessible to the large amounts of acetylcholine molecules that accumulate due to poisoning. Atropine blocks the muscarinic receptors at parasympathetic effector organs and in the central nervous system. Oxine-2 PAM (2-pyridine aldoxine methiodide) is a specific antidote for organophosphate poisoning. Aminophylline potentiates organophosphate toxicity and should not be used. The use of 2-PAM combined with atropine is valuable in organophosphate poisoning. Pyridine Aldoxine Methiodide reactivates Acetylcholinesterase in the body. It should be given when there is full antropinization.

Chapter 3
Ecotoxicology

The term ecotoxicology was introduced by Trauhaut in 1969 and this field is a natural extension of toxicology. It is best defined as the study of the fate and effects of toxic substances on an ecosystem and is based on scientific research employing both field and laboratory methods. Environmental toxicology as it is related to ecotoxicology requires an understanding of ecologic principles and theory as well as a grasp of how chemicals can affect individuals, populations, communities and ecosystems. Ecotoxicology builds on the science of toxicology and the principles of toxicologic testing, though its emphasis is more at the population, community and ecosystem levels.

Descriptions of ecotoxicologic methods and procedures have been offered by Cairns (1978) and Cairns et. al (1980) and more so by Hoffman et. al (1975). Unlike standard toxicologic tests, which seek to define the cause effect relationship with certain concentrations of toxicant exposure at a sensitive receptor site, ecotoxicologic testing attempts to evaluate cause and effects at higher levels of organization, but particularly on populations. To a large extent, the early tests (such as evaluating the effects of pesticides in fish and wildlife populations) generally employed species-specific tests in the laboratory. Tests of species included aquatic species such as *Daphnia magna*, fathead minnows (*Pimephales promelas*), the mosquito fish (*Gambusia affinis*) and among wildlife the northern bobwhite (*Celosias virginians*) and mallard duck (*Anas platyrhynas*).

A critical component of ecotoxicologic testing is the integration of laboratory and field research. Laboratory toxicity bioassays define toxicant impact on industrial organisms and on their biochemistry and physiology. Laboratory testing often limits the complexity of stress parameter except perhaps for isolating the toxicant. It is therefore difficult to interpret potential ecotoxicologic effects resulting from laboratory studies without data from pertinent field investigations. For these reasons, integrating laboratory and field research ensures that the ecotoxicologic testing method produce relevant data. Demands on ecotoxicologic testing methodologies will continue to increase as a concern for environmental protection and chemical impacts are increasing. Furthermore, there is an increasing interest in the relationship of the environment and potential environment toxicant stressors in human health implications. In recent years, the creation of major new environmental legislation-including the Food Quality Protection Act of 1966 and amendments to the Safe Drinking Water Act has dictated a renewed evaluation of the relationship between environmental toxicants and potential impacts of the environment as related to human health implications.

Chemical Movement, Fate and Exposure

To characterize chemical behaviour, it is necessary to measure the chemical in different environmental compartments (e.g. air, soil, water and biological systems), understand the movement and transport of the chemical within and among these compartments, and follow the chemical as it is metabolized, degraded, stored or concentrated within each compartment. During the past half-century, intensive efforts has been directed towards developing analytic techniques to detect and quantify minute concentrations of chemicals in environmental matrices. Consider, for illustrative purposes, the fact that 1ppt is 1 billion times smaller than a part per million (ppm), equating to approximately 1.0g of salt in a billion metric toxic of sugar. Nevertheless it is well documented that environmental concentrations below 1.0ppm of certain chemicals can have deleterious effects on different components of the ecosystems.

Chemodynamics

Chemical transport occurs within environmental compartments (interphase) and between them interphase and is critical in understanding and interpreting environmental toxicology data. Chemodynamics is an essence, the study of chemical release, distribution, degradation, and fate in the environment. Chemodynamics can also describe chemical movement or absorption into the organisms. Detoxification mechanisms such as partitioning into adipose tissue, metabolism and accelerated excretion can significantly reduce eliminate, or in some cases increase the toxic action of the chemical. Thus chemodynamics aids in prediction of chemical concentrations in compartments, serves as a source for designing toxicologic experiments using the appropriate concentrations and forms of the chemical in question.

Single-Phase Chemical Behaviour

Once a synthetic chemical enters the environment, it is acted upon primarily by natural forces. Models are used to predict the effect of natural forces on the movement of chemicals in the environment. This requires the incorporation of biotic variables into valid models. These variables include temperature, wind and water flow directions and velocities, incident solar radiation, atmospheric pressure and humidity and the concentration of the chemical in one of the four matrices: atmosphere (air), hydrosphere (water), lithosphere (soil) and biosphere (living organisms). Intraphase movement consists of mass transfer, diffusion, or dispersion within a single phase. Concentration gradients result in movement within the medium. Contaminant persistence is a function of the stability of that chemical in a phase and its transport within that phase.

Air: The primary routes of contaminant entry into the atmosphere are through evaporation, stack emissions and other matrices. Contaminant transport in air generally occurs much more rapidly than in the hydrosphere, as air has lower viscosity. Contaminant transport in air occurs primarily by diffusional processes or advection. Diffusion dominates in the very thin boundary layer between air and other phases, the thickness of which is less than that of equivalent water-phase interfaces. The diffusion rate for a contaminant in air is approximately 100 fold faster than for the same contaminant in water and in function of phase viscosity and existing concentration gradients. The contaminant diffusivity in air depends on its molecular weight compared to air, air temperature, the molecular collision, the energy of molecular interaction and Boltzmann's constant. Wind currents transport airborne contaminants much more rapidly than does diffusion. Atmospheric stability affects the amount of turbulence and thus the degree of vertical mixing in the atmosphere. The stability of the atmosphere is considered neutral when the convective forces–heat transfer from warm ground surface and radioactive cooling from the top of the cloud layer–are equal. Vertical mixing is at maximum when heat transfer is greater than radioactive cooling and at a minimum during inversion conditions. It is the latter condition that can trap higher concentrations of contaminants near the earth's surface.

Water: Contaminants enter the hydrosphere by direct application spills, wet and dry deposition, and interphase movement. In addition, chemicals enter the hydrosphere by direct dissolution of lighter than-water spills in the form of slicks or from pools on the bottom of channels, rivers or other waterways. Chemical movement in the hydrosphere occurs through diffusion, dispersion and bulk flow of the water. In any flow, a stagnant boundary layer exists at the interface between phases or artificial boundaries. Overlying this layer is a section in which flow is laminar. Finally above the laminar flow, the fluid is in turbulence flow. Contaminant movement in the mobile phase is dominated by the turbulence of the mobile phase. If the water is stagnant, the chemical moves by molecular diffusion. As described for the other fluid environmental compartment, air, the diffusion rate depends on fixed characteristics such as the molecular weight of the contaminant (solute), the molecular weight of the contaminant (solvent), water temperature, viscosity and the association factor for water and dynamic characteristics such as the magnitude of the concentration gradient of the contaminant. These characteristics are referred to as the diffusivity of the contaminant–water mixture.

Soil: Chemicals entering the lithosphere by processes similar to those for the hydrosphere soils have varying porosities due to their composition (percent sand, silt, clay, organic matter), but pores are filled with either gas or fluids. Chemical movement in the soil occurs by diffusion in these fluids or by movement of water through the voids between soil particles. The direction of diffusion will be from areas of high to areas of low concentration.

Chemical Behaviour and Bioavailability

An appreciation of how physico-chemical properties influence contaminant behavior is necessary to anticipate chemical concentrations and speciation of different environmental compartments.

In the environment, only a position of the total quantity of chemical present is potentially available for uptake by organisms. This concept is referred to as the biological availability of a chemical. Chemical bioavailability in various environmental compartments ultimately dictates toxicity; therefore it is important to characterize exposure on a site specific basis. For example, total mercury concentration in aquatic sediments does not necessarily correlate with mercury concentration in midge larvae of the genus *Chironomus*. In most cases, mercury will not exist as a single species but will be distributed among several stable forms. Hence, a simple analytic result of total mercury content does not have sufficient desirable association with the metal in the sediment.

The behavior and bioavailability of sediment-incorporated xenobiotics is a complex phenomenon studied only recently. The awareness that many aquatic contaminants settle into sediments has prompted studies of metals and organics to characterize their fate and deposition within the complex sediment matrix. Deposition is a combination of physical, chemical and biological processes that may ultimately change the form of xenobiotic. Many metals are abiotically or biotically reduced as they are incorporated into the sediments.

Biomarkers

A fundamental challenge in environmental toxicology is relating the presence of a chemical in the environment with a valid prediction of ensuring hazard to biological receptors. Adverse health effects in biological receptors begin with exposure to a contaminant and can progress to damage or altera in function of an organelle, cell or tissue. Exposure of wildlife by contact to contaminated environmental media is defined as an external dose, whereas internalization of the contaminated media, via inhalation, ingestion or dermal absorption results in internal dose. The amount of the internal dose necessary to elicit a response or health effect is referred to as the biological effective dose.

The National Academy of Sciences defines a biomarker or biological marker as a xenobiotically induced alteration in cellular or biochemical components or processes, structures or functions that is measurable in a biological system or sample. Therefore, biomarkers can be broadly categorized as markers of exposure, effects or susceptibility. The selection of appropriate biomarkers to be used for hazard evaluation is based on the mechanism of a chemically induced disease state. Moreover, growing awareness of the possibility of using wildlife as sentinels for human environmental diseases has created a demand for biomarkers that are non-lethal and correlate with adverse effects in humans.

Dosing with an adequate concentration of a toxicant produces a continuum of responses beginning with exposure and perhaps resulting in the development of disease. These events begin with external exposure, followed by the establishment of an internal dose leading to delivery of a contaminant to a critical site. This is followed either by reversible or irreversible adverse alterations to the critical site, resulting in development of recognizable disease states. A clearer understanding of the disease state in a species leads to an increase in the number of specific and useful biomarkers that may be extrapolated to other species.

Biomarkers of Exposure

The presence of a xenobiotic substance or its metabolite(s) or the product of interaction between a xenobiotic agent and some target molecule or cell that is measured within a compartment of an organism can be classified as a biomarker of exposure. In general, biomarkers of exposure are used to predict the dose received by an individual, which can be related to the changes resulting in the disease state. In many cases, biomarkers of exposure are among the most convenient to determine because the contaminant or its metabolites can be quantified from nonlethally obtained samples of exhaled air, urine, faeces, blood or breast milk as well as tissues obtained through biopsy or necropsy.

Some useful biomarkers of cancer are involved in detecting the ability of chemical carcinogens to form adducts with cellular macromolecules such as DNA or protein. Most chemical carcinogens are either strong electrophiles or are converted into an electrophillically active substance through metabolic activation.

Biomarkers of Effect

Biomarkers of effect are defined as measurable biochemical, physiologic, behavioral or other alterations within organisms that within an organism depending on the magnitude can be recognized as established or potential health impairment or disease. Ideally, a biomarker result must be able to stand alone. As such, it does not need chemical analysis or additional biological tests for confirmation. These tests are highly specific for individual chemicals and thus have a fairly limited application. Examples of such biomarkers include inhibition of brain cholinesterase by organophosphate or carbamate insecticides.

Less specific biomarkers are also well validated, but they have wider applications and tend to respond to broader classes of chemicals. Examples of these biomarkers are the induction of mixed function oxidizes, the formation of DNA adducts, other DNA alterations such as sister chromatid exchange and strained breakage, porphyrin profile alterations, induction of vitellogenin in oviparous vertebrates and immunologic changes such as immunosuppression and hypersensitivity. These assays require either additional biomarker studies or chemical residue analysis in order to link causative agent to adverse effects. For example, induction of cytochrome P4501A1 (CYP1A1) enzymes in fish liver is generally recognized as a useful biomarker of the exposure of fish to anthropogenic contaminants, but these results are not compound-specific as they may be induced by a variety of polynuclear and halogenated aromatic hydrocarbons as well as by hypoxia.

Finally, there is a long test of biomarker that are under development or have been used with varying degrees of success but require further validation before they can be used in hazard evaluation. Thyroid function, retinol levels, plasma sex steroids and stress proteins fall into this classification.

Biomarkers of Susceptibility

Biomarkers of susceptibility are end points that are indicative of an altered physiologic or biochemical state that may predispose the individuals to impact of chemical, physical or infectious agents. These biomarkers can be useful in predicting human disease states from wildlife sentinels. Low-level exposure to a cytochrome P450 A1 or A2 inducer, for example, may result in elevated enzyme activity in wildlife but no observable adverse effects. Such elevations in enzyme activity in humans have been linked to greater risk of a number of cancers due to increased bioactivities of pro-carcinogens.

Biomarker Interpretation

Caution must be used in interpreting biomarker results and extrapolating from one species to another. The same chemical may induce different proteins in one species when compared to another and the same enzyme may have different substrate specificities in species as closely related as the mouse and rat. For example, the common environmental contaminant p,p^1-DDE induces cytochrome P4502B in the laboratory rat (Rattus rattus) but induces cytochrome P4501A1 in the deer mouse (*Peromyscus maniculatus*).

Alternatively, an important application of biomarkers is their ability to integrate multiple chemical exposures across an area with a variety of chemical contaminants, the scenario found at most chemical waste sites.

Beyond the current predominance area of functionally-based biomarkers, new trend in biomarker development appear distinctly molecular. A review of the most recent biomarker literature lists molecular biomarkers for a great many diseases and environmental contaminants ranging from second hand smoke to suicide. The integration of biomarkers with epidemiology has resulted in a new discipline, molecular epidemiology, which has the potential for creating worldwide databases for environmental and genetic diseases. The integration of biomarkers with molecular biology has both revolutionized medicine and biology by providing new tools to determine the mechanism of action. Moreover, these techniques can be applied to samples as small as one cell. Increasing emphasis is being placed upon non-lethal biomarkers such as enzyme linked immune sorbent assay (ELISA) techniques for measuring fecal steroids in deer mice. New biomarkers are needed to monitor the continuum between exposure and overt disease. An increased knowledge is needed for the relationship between biomarker responses and disease pathology.

Endocrine and Developmental Disruptors

Endocrine disruption has recently emerged as a major issue in terms of science and public policy. A number of compounds, both natural and anthropogenic, cause alterations of the endocrine system.

Mechanisms of Endocrine Toxicity and Sensitive Life Stages

It is evident that endocrine disrupting compounds (EDCs) may interact with multiple targets. There is evidence for EDCs acting at every level of hormone synthesis, secretion, transport, site of action and metabolism. Some examples of known mechanisms for EDCs include the following:-

Receptor-mediated effects of EDCs: A xenobiotic compound may exert effects of the receptor level through multiple mechanisms beyond the classic legend-receptor interaction. These include differential effects at multiple receptor sites or direct effects on intercellular signaling pathways, thereby directly influencing hormone action at the target tissue. Xenobiotic compounds may act on endocrine system by affecting transcription and signal transduction and can act through receptor-mediated or non receptor-mediated mechanisms. For example, genistein has been shown to be a week estrogen receptor agonist, however it also modulates the activity of tyrosine kinesis and DNA topoisomerases.

Effect of EDCs on hormone synthesis and metabolism: A compound may adversely alter levels of critical endogenous hormones by inducing or inhibiting biosynthetic or metabolic enzyme activities. Some phytoestrogens can interact with the 17β dehydrogenase that regulates estradiol and estrogen levels, suggesting that they can modulate overall estrogen levels in addition to acting as a legend for the estrogen receptor.

Effects on hormone secretion and transport: It has been known for many years that Cd^{2+} is a non-selective Ca^{2+} blocker that can disrupt Ca^{2+} dependent exocytosis in hypothalamic neurosecretory neurons and pituitary endocrine cells. Alternatively, EDCs can affect hormone-binding [sex hormone binding globulin, (SHBG), corticosteroid binding globulin (CBG)], proteins in blood, thereby disrupting hormone transport by increasing or decreasing the bound-to free ratio of the hormone in plasma.

Timing of exposure: There is substantial evidence that the sensitivity of an individual to gonadal steroids is dependent on the life stage of the individual. Specifically, the foetus appears to be the most sensitive life stage for lasting impacts of gonadal steroids or agonists/antagonists.

Hormone regulation and feedback control: There are several important control mechanisms that regulate estrogen biosynthesis during pregnancy. Estrogen levels are not feedback regulated in a typical homeostatic mechanism, rather there is feed forward mechanism resulting in steadily increasing serum levels of estradiol across most of the pregnancy in rodents and humans. Thus an exogenous dose of any estrogen agonist will be additive with the exogenous level because feedback will not reduce exogenous production in a compensatory way.

Species dependent sex determination: There are major differences in the control of sex determination among vertebrate classes. In mammals, sexual determination is based on XY/XX system with the female as the homogametic sex. This system requires the synthesis of testosterone and dihydrotestosterone in some target tissues and the presence of functional androgen receptors' in the undifferentiated gonad, secondary sexual tissues and brain. In rodents the presence of estrogen receptors in the brain occur, testosterone or a precursor must be aromatized to 17β-estradiol.

Terrestrial and Aquatic Ecotoxicology

Many environmental studies include the analysis of contaminant exposure and effects on relatively small scales. However, contaminants can affect ecologic systems over large areas, including ecosystems and landscapes. Ecosystems are composed of groups of all types of organisms that function together as well as interact with the physical environment, including energy flow and cycling of materials between living and non-living components. In turn, ecosystems collectively constitute landscapes with their own functional (nutrient and energy flow) and structural (patches, corridors) attributes. Movement of biotic and biotic components within these large systems varies depending on several factors, including the species of animal and physical features of the system. Large vertebrates may roam over hundreds of square kilometers integrating many habitat types within their home range. The area used by small animals may be small on an individual basis; however, dispersal individual can maintain rather extensive connectivity among otherwise distinct local populations. In general, ecologic systems are in a constant state of communication, which can potentially facilitate the large-scale effects of pollution.

Ecotoxicology includes all aspects of aquatic and terrestrial systems while attempting to elucidate the effects of biota following contaminant exposure. Studies in aquatic and terrestrial toxicology rely heavily on interdisciplinary scientific exploration. Such research encompasses a variety of topics, including toxicity testing, sublethal responses of individual organisms, effects on populations and communities and field research. These biological indicators of population may include individual-based measurements of some biochemical, physiologic or morphologic endpoints or higher endpoints measurements including perturbations of populations at higher levels.

Separation of aquatic and terrestrial environments in ecotoxicology is often impossible, as contaminants can be readily transported between these two systems. For example, contaminants in terrestrial environments may be transported to aquatic organisms' located considerable distances from the source of contamination. Conversely, contaminants originating in aquatic environments may move into terrestrial environments following flood events of evaporation. One mechanism of contaminant movement of considerable interest in ecotoxicology is the transfer of contaminants through trophic levels, both within and between aquatic and terrestrial systems.

Toxicity tests: Acute and chronic toxicity tests are designed to determine the short and long term effects of chemical exposure on a variety of endpoints, including survival, reproduction and physiologic and biochemical responses. Toxicity testing of terrestrial animal and plant species serves a number of purposes in terrestrial toxicology. Advances in toxicology have resulted in an expanding search for new sentinel plant and animal species for assessing contaminant exposure and effects. Acute and chronic toxicity testing represents the initial steps towards validating new animal and plant species as useful sentinels of environmental contamination.

Sublethal effects: Mortality represents a non-reversible endpoint of interest in ecotoxicology. The existence of sublethal effects in exposed organisms has been used as an advantage in monitoring strategies. Inhibition of cholinesterases has proven an excellent marker that is both sensitive and diagnostic for organophosphate and carbonate insecticide exposure. Induction of enzyme systems such as mixed function oxygenases are also useful as sublethal biomarkers of exposure to many types of environmental pollutants. Other strategies for monitoring sublethal effects include monitoring immune function and reproductive endpoints. Even though these effects may not result in immediate mortality, they can effect fecundity and reproductive success of aquatic and terrestrial organisms and ultimately have effects on population structure and function. Chemicals may also affect the growth rate of organisms. Because growth rate and body size are related to reproductive maturity in juvenile organisms and its attainment as well as relative fecundity in adults, chemical stressors that inhibit growth rates can also affect the reproductive potential of the population.

Determination of sublethal effects is an important component of risk assessments for two reasons. **Firstly** these responses may provide information not available from measurements of contaminant tissue concentrations. This is because:- (1) It may be possible to measure tissue concentrations of some chemicals because they are rapidly metabolized. (2) The toxic effects of many chemicals, especially when present in complex mixtures may not be predicted from tissue concentration alone.

Secondly alterations of biochemical and molecular physiology has been associated with reductions of fecundity, growth and bioenergetics status of affected organisms. Hence, perturbations of subcellular function may affect fitness and health of fish and wildlife and ultimately may be translated as effects on populations and communities.

Although they are quite similar but sublethal effects in aquatic and terrestrial organisms do differ in some important aspects. For example, there are number of suborganismal effects that can be detected in aquatic and terrestrial organisms, and many of these are commonly studied in both types of organisms e.g., liver mixed function oxidase induction. However, aquatic and terrestrial organisms may differ in the relative magnitude of these responses. DNA-repair enzyme activity may be lower in fish than in mammals. There are also differences between aquatic and terrestrial biomarkers in relation to the attention given to at various endpoints. For example, studies that examine acetylcholinesterase inhibition mainly focuses on terrestrial organisms, whereas studies examining DNA damage are more heavily represented in aquatic studies. Another class of protein that can be induced by contaminant exposure is the stress or "heat shock" proteins which participate in the renaturation of damaged proteins. Aquatic toxicology studies also differ from those in terrestrial toxicology because aquatic organisms respire through gills. Gills may be exposed to water-borne contaminants and are highly permeable to dissolved substances. As a result, gills may accumulate certain contaminants or their structure and function may be impaired.

Population and community effects: One of the major objectives of ecotoxicology is the detection and prevention of pollutant effects on population structure and function. The age structure of populations may give an indication of population effect, such as reproductive failure or perturbations in recruitment of juveniles into the population. The pattern of population response to pollution may provide information as to the mechanism of population effect, such as changes in adult mortality, juvenile recruitment, food availability, etc.

Any effects on populations may ultimately manifest its effects on communities, because by definition, communities are collections of interacting populations. Environmental contaminants can affect the structure of communities as well as the interactions of species within them. For example, it is well known that exposure to chemicals may cause a reduction of community diversity and changes in community composition.

The trophic structure of communities is related to the relative abundance of species that feed on various food items or have various foraging methods. These changes in trophic/species composition may come about by direct or indirect mechanisms. The direct effects involve loss of some species due to increase in pollution induced mortality or reduced reproductive output. In this case the communities will be dominated by species that are less affected by pollutant exposure. This is the basis of phenomenon termed pollution-induced community tolerance or PICT, in which the algal communities become more pollution-tolerant over time due to replacement of pollution-sensitive species with more tolerant ones.

Alternatively, community structure may change through indirect mechanism. For example, a species may be absent from a community because the organisms upon which it feeds are exterminated by pollutant exposure. Indirect effects may also be affected by changes in dynamic interactions between species, for example, predator/prey interactions. Environmental contamination may give one species a competitive edge over the other, resulting in local extinction of less tolerant species. Finally, it has been suggested that such changes in community structure came about because some species are more genetically adaptable than others and so are better able to adapt to novel stressors such as pollution. Thus, the more sensitive species would not be able to adapt to this stressor and become locally extinct. These types of perturbations in community structure and dynamics may ultimately compromise the stability, sustainability and productivity of the affected ecosystem.

Chemical Interactions and Natural Stressor

As more information becomes available on chemical effects in aquatic and terrestrial organism, there is an increasing interest in understanding the interaction effects of exposure to multiple contaminants as well as the interaction between contaminants and inherent stressors (e.g. nutritional stress, disease, predation, climate, water quality). Perhaps, the greatest inherent stressors faced by many species of wildlife are nutritional restrictions and seasonal shifts in climate extremes. Daily food restriction of as little as ten percent below normal intake has been shown to enhance the overall decline in courtship behaviour, egg laying, hatching, and number of young fledged by rigned turtle doves exposed to DDE. Unlike many wildlife species, fish and aquatic invertebrates are poikilotherms, so their metabolic rate is more dependent on ambient temperature than that of birds or mammals. As a result, toxicity accumulations and metabolism of aquatic contaminants may be influenced by water temperature. Other environmental variables, such as salinity and pH may also affect uptake and toxicity of aqueous chemicals.

Trophic-Level Transfer of Contaminants

Although, contaminant exposure may occur through inhalation, dermal contactor indigestion from preening or grooming behavior, significant exposure also occurs through food chain transport. Depending on specific chemical properties, contaminants may become exposed through ingestion of contaminated prey, promoting accumulation or magnification of contaminants into higher trophic levels. Earthworms in soils contaminated with organochlorines and heavy metals can accumulate quantities of contaminants known as the deleterious to sensitive species. The use of pesticides to control plant pests often coincides with the reproductive periods of many wildlife species, enhancing exposure potential in juveniles that relay on invertebrates as a primary food source.

Genotoxicity

Ecogenotoxicology is a relatively emerging area that has benefited tremendously from the growth of molecular biology and molecular genetics. It is concerned with the effects of pollutants or chemical (mutagens, clastogens, aneuogens and teratogens) on the genetic material of the organisms. Such genetic material is usually defined as DNA, RNA and chromosomes but may also include modifications of proteins. Such effects may manifest as DNA strand breaks, base modifications, and chromosomal rearrangements or fragmentation and aeneuploidy. While it is possible to damage the genetic material of an organism without any subsequent effect on the individual, it is also possible that mutations in the DNA can result in somatic effects such as cancers. If these effects occur in germinal tissues, this can result in heritable effects and an increase in genetic load. They proposed that the selection resulting from stress of somatic effects of contaminants could lead to population genetic changes that are predictable from the knowledge of the mechanism of toxicology of contaminants.

Besides selection and genetic bottlenecks, an elevated mutation rate may also alter population genetic structure. The mitochondrial DNA has the least effective repair mechanism and should be among the fragments of DNA that permit detection of an elevated mutation rate.

Terrestrial Ecotoxicology

Terrestrial toxicology is the science of the exposure to and effects of toxic compounds in terrestrial ecosystems. All organisms function at several levels, from the individual level of the ecosystem, interacting with others within the constraints of social ranking, food webs and niches. Many terrestrial species are very mobile, covering significant areas while defending territories, foraging, migrating and dispersing. It includes all aspects of terrestrial system while attempting to elucidate the effects on biota following contaminant exposure. Like aquatic toxicology, the terrestrial toxicology relies on interdisciplinary scientific exploration.

The early 1900s witnessed the realization that chemicals used in the environment could affect non-target organisms. In latter years, synthetic pesticides became increasingly important in controlling pest species in agricultural crops, although little was known about their effects on non-target organisms.

Chemical effects on avian population were the primary focus for many years. The problem became more apparent as the link was established between DDT contamination and declining bird populations.

Acute and chronic toxicity testing: Terrestrial organisms are typically exposed to contaminants through ingestion of some contaminated media, although inhalation and dermal absorption of contaminants do occur. Thus, toxicity tests for terrestrial species are usually designed to test the effects of a chemical dose, administrated by oral gavages or injection. Exposure can also be accomplished through consumption of contaminated food or water, resulting in dosages calculated from consumption rates or simply exposure over time to a given concentration of contaminant in the diet. Methods for measuring endpoints in toxicity tests include the LD_{50} and LC_{50}, the ED_{50} and EC_{50} and reproductive tests. These endpoints can be used to assess toxicity in a variety of terrestrial animals, including earthworms, honeybees, northern bobwhite, mallards, mink and European ferrets. Likewise, specialized tests for determining toxicity in plants are used to assess lethal and non-lethal response to contaminants.

Field testing: Field studies are designed to address exposure to contaminants and resulting effects to organisms outside the highly controlled environment of the laboratory. Field studies are conducted in complex ecologic systems where plants and animals are affected by numerous natural stresses (e.g. nutrient restriction, disease and predation) that might possibly confine measurement of contaminant exposure and effects.

Traditional methods used by biologists have been used with success in terrestrial ecotoxicology field studies and resources are available that describe the various techniques for trapping, remote sensing and sampling terrestrial biota.

Aquatic Ecotoxicology

Aquatic ecotoxicology is the study of effects of anthropogenic chemicals on organisms in the aquatic environment. The aquatic ecosystem is of particular concern because this is where most contaminants released into the environment are eventually deposited, either from direct discharge into the bodies of water or from terrestrial run-off and atmospheric deposition. There are certain features of the aquatic environment that make it unique. First, certain chemicals are not volatile in the air but are soluble in water (e.g. metals), so aquatic organisms may be exposed to chemicals via routes that are not present in terrestrial counterparts. Also, many contaminants are readily degraded in the aerobic environment, but the aquatic environment frequently contains little or no oxygen. Therefore, some contaminants can persist in aquatic ecosystems far longer than the terrestrial systems. Finally, the aquatic organisms are frequently restricted in their habitat and home range, so that they often cannot avoid contaminated areas.

The physiology and anatomy of aquatic organisms can also present problems that are different from those faced by terrestrial organisms. For example, aquatic organisms have highly permeable skin and gills and so are particularly susceptible to the effects of ambient contamination. The aquatic communities are dominated by ectothermic organisms (e.g. invertebrates, fish, amphibians and to a lesser extent, aquatic reptiles), whose metabolic rate is determined by ambient water temperature. Thus the accumulation of contaminants and their toxic effects are influenced by water temperature, which can vary both spatially and temporally.

Aquatic and chronic toxicity testing: In aquatic toxicity tests, fish, invertebrates or algae are exposed to aqueous chemicals in the laboratory. Designs for these tests include static, static renewal and flow through systems. The organisms may be exposed to short or long periods. Acute and chronic tests not only differ in duration but also in the endpoints that are measured. However in chronic tests, effects on growth and reproduction are also determined. To accomplish this, the duration of chronic test is to span the life cycle of the organism. However, chronic tests are often difficult to perform because of very long duration and they are expensive, which makes their routine use prohibitive.

Another type of test used for monitoring purposes is sediment toxicity testing. In this case sediment is collected from the bottom of lakes, rivers, bays, etc. and brought to the laboratory. Important application of sediment tests include determination of toxicity of sediments that are dredged from one location (e.g. clearing shipping channels) and need to be disposed-off at another locations, environmental risk assessments of contaminated areas, and as part of bio-monitoring programs for the purpose of compliance to environmental regulation.

Sub-lethal effects: In the aquatic environment, concentrations of contaminants in the water or sediment may not be high enough to elicit mortality but may still induce sub-lethal effects on the health of aquatic organisms. One method by which these health effects can be determined is via evidence of tissue damage or dysfunction. This type of damage can lead to diseases such as tumor or infectious and parasitic infestations.

Field studies: Aquatic field studies are either manipulative or observational. In manipulative studies, previously unexposed organisms are used, and the experiments determine the level of contamination to which they are exposed.

Good Laboratory Practices in Terrestrial and Aquatic Ecotoxicology

Good scientific practices, which results in high quality data collection and interpretation are of paramount importance in the field of toxicology. The GLPs are regulation standards that define conditions, under which a toxicology study should be planned, conducted, monitored, reported and archived. They have been adopted by many national and international governments and agencies. The GLPs outline study management procedures and documentation practices that, if followed, will limit the influence of extraneous factors on study results and interpretations. GLPs include provisions for such factors as personnel management and training facilities, support and operation, equipment design, maintenance and calibration, independent quality assurance monitoring, handling of test systems and materials, documentation of study conduction, written standard operation procedures and study protocols, study records and retention of records and samples.

Modeling and Geographic Information Systems

Modeling in ecotoxicology allows the prediction of effects of toxic compounds on the environment, which can be characterized by various ecosystems. Terrestrial ecosystems include forests, grasslands and agricultural areas, whereas aquatic ecosystem includes lakes, rivers and wetlands. Each of these systems is a collection of interconnected components, or subsystems that function as a complete entity. Because the dynamics of real systems are quite complex, an understanding of the impacts of toxicants on a system can be enhanced by modeling that system.

The modeling process involves two steps (1) Identification of component interactions (2) Characterization of those interactions using quantitative abstraction of mechanism processes. Once the model has been defined, it is implemented on the computer. Measurements obtained from the real system are compared with model projections in a process of model validation. Improvements are then made to the model by changing parameter values or modifying equations in the model. Several interactions of comparing model behavior to that of the real system are usually required to obtain a satisfactory or "valid" model.

Types of Models

Models can be used to obtain qualitative or quantitative information about a complex system. Quantitative models emphasize the relationship between the variables of interest while minimizing the requirement of tremendous accuracy in the parameter of the model. The disadvantage is that there can be no reliance on the number produced from such a model. Quantitative ecotoxicologic models can be classified as (1) individual based versus aggregated models (2) stochastic versus deterministic models (3) spatially distributed versus lumped models. Such models often have hundreds of thousands of lines of codes and are used only at the final stages of system stimulation.

Individual based versus aggregated models: Models that stimulate all individuals simultaneously are referred to as individual based models. Each individual in the stimulation has a unique set of characteristics: age, size, condition, social status and location in the landscape as well as its own history of daily foraging, reproduction and mortality. The individual based approach has several advantages. It enables the modeler to include complex behaviour and decision making by individual organisms in the model. But importantly, it allows one to model populations in complex landscapes, where different individuals may be exposed to very different levels of toxicant concentration.

Models of individuals can be extended to a population as a whole by:- (1) stimulating not just one individual that make up the population of interest (2) aggregating various populations' members into classes, such as age classes. Aggregated models, then, follow not individual organisms but variables representing the numbers of individual per age class. In stimulating a complex environmental system, both individual based and aggregated models will be needed. Usually individual based models are used to represent vertebrate species while aggregated models are used to represent organisms at lower classification levels.

Stochastic versus deterministic models: Model coefficients can be functional not only of other variables but also of random variables, thus they can be random variables themselves. In this way of classifying models, those with random variables are called stochastic models and those without are called deterministic models. Random variables are used to represent the random variation or "unexplained" variation in the state variables. Stochastic models also can include random variables expressed either as random inputs or as parameters with a random error term.

Ecologic Risk Assessment

With the growth of environmental toxicology, comes the need to assess and quantify the impact of toxic chemicals on organisms, their populations and communities in ecosystems. Earlier techniques to conduct risk assessments utilizing human health approaches were not appropriate for ecologic systems. This framework, which was expanded and modified in 1988, allows for the assessment of the impact of toxic chemicals as well as other stressors on ecologic systems. As part of the problem-formulation phase, a conceptual model is usually developed describing routes of exposure, biota or concern and anticipated effect endpoints. The actual risks of chemicals to wildlife and other biota is then determined using exposure data and toxic effects of the chemicals of interest.

Environmental Toxicology and Human Health

Links between wildlife and human health serve as a premise for extrapolation in risk assessment. Humans share many cellular and subcellular mechanisms with wildlife species. Humans and wildlife also overlap in their physical environment and therefore are exposed to many of the same contaminants. There is evidence to suggest that when highly conserved systems are targeted by environmental toxicants, both ecosystems and human health suffer.

There are obvious challenges and concerns in the extrapolation of wildlife data on humans. When there are contaminant-specific alterations in wildlife health, concerns about co-ordinate adverse effects in humans tend to focus on susceptible developmental periods including in utero, neonatal, pubertal, lactation and menopausal stages. There is also a real concern about an increased risk of various cancers caused by environmental contaminants and populations with genetic or environmental susceptibility. The overall rate of some cancers is increasing particularly in the industrialized countries. Based on animal models, chemical exposure figures in the etiology of many cancers, therefore a link to human cancer incidence seems plausible. Unfortunately, linking known contaminant exposures to an affected human population is difficult, particularly when effects are not identified for many years.

Regardless of species, the process of risk assessment requires four steps, hazard identification, dose response assessment, exposure assessment and risk characterization. Often these processes are difficult in human populations and extrapolations are required, including quantitative interspecies extrapolation from test animal to human and quantitative extrapolations have sometimes resulted in a low confidence in risk estimates for humans. Obviously, the more human data available for risk assessment, the better and more generalized and relevant it is to real human health effects, the easier it is to define a risk management strategy.

The interconnections between ecologic health and human health should not be overlooked. The indirect effects of environmental pollution may in the end, be more important than the direct effects for human health. The environment is thought to act as a buffer for both the toxicants and disease. However, even the buffer has its limits. For instance, the human population is at a greater risk for emerging diseases as the natural environment dwindles in relative area.

Chapter 4

Occupational Safety and Health

It is a cross-disciplinary area concerned with protecting the safety, health and welfare of people engaged in work or employment. As a secondary effect, it may also protect co-workers, family members, employers, customers, suppliers, nearby communities, and other members of the public who are impacted by the workplace environment. It may involve interactions among many subject areas, including occupational medicine, occupational (or industrial) hygiene, public health, safety engineering, chemistry, health physics, ergonomics, toxicology, epidemiology, environmental health, industrial relations, public policy, sociology, and occupational health psychology.

Since 1950, the International Labour Organization (ILO) and the World Health Organization (WHO) have shared a common definition of occupational health. It was adopted by the Joint ILO/WHO Committee on Occupational Health at its first session in 1950 and revised at its twelfth session in 1995. The definition reads: "Occupational health should aim at: the promotion and maintenance of the highest degree of physical, mental and social well-being of workers in all occupations; the prevention amongst workers of departures from health caused by their working conditions; the protection of workers in their employment from risks resulting from factors adverse to health; the placing and maintenance of the worker in an occupational environment adapted to his physiological and psychological capabilities; and, to summarize, the adaptation of work to man and of each man to his job."

The reasons for establishing good occupational safety and health standards are frequently identified as:

- **Moral** - An employee should not have to risk injury or death at work, nor should others associated with the work environment.
- **Economic** - Many governments realize that poor occupational safety and health performance results in cost to the State (e.g. through social security payments to the incapacitated, costs for medical treatment, and the loss of the "employability" of the worker). Employing organizations also sustain costs in the event of an incident at work (such as legal fees, fines, compensatory damages, investigation time, lost production, lost goodwill from the workforce, from customers and from the wider community).
- **Legal** - Occupational safety and health requirements may be reinforced in civil law and/or criminal law; it is accepted that without the extra "encouragement" of potential regulatory action or litigation, many organizations would not act upon their implied moral obligations.

INDIAN SCENARIO

Hazards, Risks, Outcomes

The terminology used in Occupational Safety and Health (OSH) varies between states, but generally speaking:

- A hazard is something that can cause harm if not controlled.
- The outcome is the harm that results from an uncontrolled hazard.
- A risk is a combination of the probability that a particular outcome will occur and the severity of the harm involved.

"Hazard", "risk", and "outcome" are used in other fields to describe e.g. environmental damage, or damage to equipment. However, in the context of OSH, "harm" generally describes the direct or indirect degradation, temporary or permanent, of the physical, mental, or social well-being of workers. For example, repetitively carrying out manual handling of heavy objects is a hazard. The outcome could be a musculoskeletal disorder (MSD) or an acute back or joint injury. The risk can be expressed numerically, (e.g. a 0.5 or 50/50 chance of the outcome occurring during a year), qualitatively as "high/medium/low", or using a more complicated classification scheme.

Risk Assessment

Modern occupational safety and health legislation usually demands that a risk assessment be carried out prior to making an intervention. The risk management requires risk to be managed to a level which is as low as is reasonably practical. This assessment should:

- Identify the hazards
- Identify all affected by the hazard
- Evaluate the risk
- Identify and prioritize the required actions.

The calculation of risk is based on the likelihood or probability of the harm being realized and the severity of the consequences. This can be expressed mathematically as a quantitative assessment (by assigning low, medium and high likelihood and severity with integers and multiplying them to obtain a risk factor, or qualitatively as a description of the circumstances by which the harm could arise. The assessment should be recorded and reviewed periodically and whenever there is a significant change to work practices. The assessment should include practical recommendations to control the risk. Once recommended controls are implemented, the risk should be re-calculated to determine whether it has been lowered to an acceptable level. Generally speaking, newly introduced controls should lower risk by one level, i.e., from high to medium or from medium to low.

Job Safety Analysis is a methodology for performing a risk assessment.

The precautionary principle is an increasingly used method for reducing potential chemical or biological OSH risks.

Common Workplace Hazard Groups

Workplace hazards are often grouped into environmental hazards, environmental agents, physical hazards, physical agents, chemical agents, biological hazards and psychosocial issues.

- Environmental hazards:
  - Asphyxiation
  - Dehydration
- Environmental agents:
  - Cold stress (hypothermia)
  - Heat stress (hyperthermia)
  - Particulate inhalation
- Physical hazards:
  - Collisions
  - Confined space
  - Slips and trips
  - Falls from height
  - Struck by objects
  - Electricity
  - Falling on a pointed object
  - The release of contained energy. Standing in line of fire.
- Physical agents:
  - Noise
  - Vibration
  - Lighting
  - Barotrauma (hypobaric/hyperbaric pressure)
  - Ionizing radiation
- Mechanical hazards:
  - Compressed air/high pressure fluids (such as cutting fluid)
  - Crushing
  - Cutting
  - Draw in
  - Entanglement

  - Equipment-related injury
  - Friction and abrasion
  - Impact
  - Moving parts
  - Shearing
  - Stabbing and puncture
  - Workplace transport
- Biological hazards:
  - Bacteria
  - Virus
  - Fungi
    - Mould
  - Blood-borne pathogens
  - Tuberculosis
- Chemical agents:
  - Acids
  - Bases
  - Heavy metals
    - Lead
  - Solvents
    - Petroleum
  - Particulates
    - Asbestos and other fine dust/fibrous materials
    - Silica
  - Fumes (noxious gases/vapors)
  - Highly-reactive chemicals
  - Fire, conflagration and explosion hazards:
    - Explosion
    - Deflagration
    - Detonation
    - Conflagration
- Psychosocial issues:
  - Work-related stress, whose causal factors include excessive working time and overwork
  - Violence from outside the organization

- Bullying, which may include emotional, verbal, and sexual harassment
- Mobbing
- Burnout

Other issues:

- Reproductive hazards
- Avoidance of musculoskeletal disorders by the employment of good ergonomic design. Fire prevention (fire protection/ fire safety) often comes within the remit of health and safety professionals as well.
- ANSI Z10
- Environment, Health and Safety
- Material safety data sheet
- Occupational Health and Safety Management Systems - OHSMS
- OHSAS 18001

Government Organizations

- Australian Safety and Compensation Council (ASCC) (Australia)
- Canadian Centre for Occupational Health and Safety (Canada)
- Congressional Office of Compliance (US)
- European Agency for Safety and Health at Work (EU)
- Health and Safety Executive (UK)
- International Labor Organization (United Nations)
- KOSHA : Korea Occupational Safety and Health Agency (South Korea)
- National Institute for Occupational Safety and Health (US)
- Occupational Safety and Health Administration (US)
- Work Safe BC formerly Workers' Compensation Board of BC (WCB) (British Columbia, Canada)
- Workplace Safety & Health Council (Singapore)
- Workplace Safety & Insurance Board (Ontario, Canada)
- Work Safe Victoria, Australia
- National Institute of Occupational Health (India)
- Indian Institute of Toxicology Research, Lucknow (India)

Laws

- Health and Safety at Work Act (UK)
- Indonesian Act No.1/1970 about Occupational Safety at Work 1970 (Indonesia)

- Occupational Safety and Health Act (US)
- Occupational Safety and Health Act 1994 (Malaysia)
- Timeline of Major U.S. Environmental and Occupational Health Regulation
- Workplace Safety and Health Act (Singapore)

Importance of Occupational Health and Safety

Work plays a central role in people's lives, since most workers spend at least eight hours a day in the workplace, whether it is on a plantation, in an office, factory, etc. Therefore, work environments should be safe and healthy. Yet this is not the case for many workers. Every day workers all over the world are faced with a multitude of health hazards, such as:

- Dusts;
- Gases;
- Noise;
- Vibration;
- Extreme temperatures.

Unfortunately some employers assume little responsibility for the protection of workers' health and safety. In fact, some employers do not even know that they have the moral and often legal responsibility to protect workers. As a result of the hazards and a lack of attention given to health and safety, work-related accidents and diseases are common in all parts of the world.

Costs of Occupational Injury/Disease

Work-related accidents or diseases are very costly and can have many serious direct and indirect effects on the lives of workers and their families. For workers some of the direct costs of an injury or illness are:

- the pain and suffering of the injury or illness;
- the loss of income;
- the possible loss of a job;
- health-care costs.

It has been estimated that the indirect costs of an accident or illness can be four to ten times greater than the direct costs, or even more. An occupational illness or accident can have so many indirect costs to workers that it is often difficult to measure them. One of the most obvious indirect costs is the human suffering caused to workers' families, which cannot be compensated with money.

The costs to employers of occupational accidents or illnesses are also estimated to be enormous. For a small business, the cost of even one accident can be a financial disaster. For employers, some of the direct costs are:

- payment for work not performed;
- medical and compensation payments;
- repair or replacement of damaged machinery and equipment;
- reduction or a temporary halt in production;
- increased training expenses and administration costs;
- possible reduction in the quality of work;
- negative effect on morale in other workers.

Some of the indirect costs for employers are:

- the injured/ill worker has to be replaced;
- a new worker has to be trained and given time to adjust;
- it takes time before the new worker is producing at the rate of the original worker;
- time must be devoted to obligatory investigations, to the writing of reports and filling out of forms;
- accidents often arouse the concern of fellow workers and influence labour relations in a negative way;
- poor health and safety conditions in the workplace can also result in poor public relations.

Overall, the costs of most work-related accidents or illnesses to workers and their families and to employers are very high.

On a national scale, the estimated costs of occupational accidents and illnesses can be as high as three to four percent of a country's gross national product. In reality, no one really knows the total costs of work-related accidents or diseases because there are a multitude of indirect costs which are difficult to measure besides the more obvious direct costs.

Health and Safety Programmes

For all of the reasons given above, it is crucial that employers, workers and unions are committed to health and safety and that:

- workplace hazards are controlled – at the source whenever possible;
- records of any exposure are maintained for many years;
- both workers and employers are informed about health and safety risks in the workplace;
- there is an active and effective health and safety committee that includes both workers and management;
- worker health and safety efforts are ongoing.

Effective workplace health and safety programmes can help to save the lives of workers by reducing hazards and their consequences. Health and safety programmes also have positive effects on both workers morale and productivity, which have important benefits. At the same time, effective programmes can save employers a great deal of money.

I. Extent of the Problem Worldwide

A. Accidents

In general, health and safety in the workplace has improved in most industrialized countries over the past 20 to 30 years. However, the situation in developing countries is relatively unclear largely because of inadequate accident and disease recognition, record-keeping and reporting mechanisms.

It is estimated that at least 250 million occupational accidents occur every year worldwide. 335,000 of these accidents are fatal (result in death). Since many countries do not have accurate record-keeping and reporting mechanisms, it can be assumed that the real figures are much higher than this. The number of fatal accidents is much higher in developing countries than in industrialized ones. This difference is primarily due to better health and safety programmes, improved first-aid and medical facilities in the industrialized countries, and active participation of workers in the decision-making process on health and safety issues. Some of the industries with the highest risk of accidents worldwide are mining, agriculture, including forestry and logging, and construction.

B. Diseases

Some occupational diseases have been recognized for many years and affect workers in different ways depending on the nature of the hazard, the route of exposure, the dose, etc. Some well known occupational diseases include:

- asbestosis (caused by asbestos, which is common in insulation, automobile brake linings, etc.);
- silicosis (caused by silica, which is common in mining, sandblasting, etc.);
- lead poisoning (caused by lead, which is common in battery plants, paint factories, etc.);
- noise-induced hearing loss (caused by noise, which is common in many workplaces, including airports and workplaces where noisy machines such as presses or drills, etc. are used).

There are also a number of potentially crippling health problems that can be associated with poor working conditions, including:

- heart disease;
- musculoskeletal disorders such as permanent back injuries or muscle disorders;
- allergies;

- reproductive problems;
- stress-related disorders.

Many developing countries report only a small number of workers affected by work-related diseases. These numbers look small for a variety of reasons that include:

- inadequate or non-existent reporting mechanisms;
- a lack of occupational health facilities;
- a lack of health care practitioners who are trained to recognize work-related diseases.

Because of these reasons and others, it is fair to assume that in reality, the numbers of workers afflicted with occupational diseases are much higher. In fact, overall, the number of cases and types of occupational diseases are increasing, not decreasing, in both developing and industrialized countries.

II. Identifying the Cause of Occupational Disease

The cause of work-related diseases is very often difficult to determine. One factor is the latency period (the fact that it may take years before the disease produces an obvious effect on the worker's health). By the time the disease is identified, it may be too late to do anything about it or to find out what hazards the worker was exposed to in the past. Other factors such as changing jobs, or personal behaviours (such as smoking tobacco or drinking alcohol) further increase the difficulty of linking workplace exposures to a disease outcome.

Although much is understood now about some occupational hazards than in the past, every year new chemicals and new technologies are being introduced which present new and often unknown hazards to both workers and the community. These new and unknown hazards present great challenges to workers, employers, educators, and scientists.

III. The Range of Hazards

There are an unlimited number of hazards that can be found in almost any workplace. There are obvious unsafe working conditions, such as unguarded machinery, slippery floors or inadequate fire precautions, but there are also a number of categories of insidious hazards (that is, those hazards that are dangerous but which may not be obvious) including:

- chemical hazards, arising from liquids, solids, dusts, fumes, vapours and gases;
- physical hazards, such as noise, vibration, unsatisfactory lighting, radiation and extreme temperatures;
- biological hazards, such as bacteria, viruses, infectious waste and infestations;

- psychological hazards resulting from stress and strain;
- hazards associated with the non-application of ergonomic principles, for example, badly designed machinery, mechanical devices and tools used by workers, improper seating and workstation design, or poorly designed work practices.

IV. Importance of Management Commitment

All levels of management must make health and safety a priority. They must communicate this by going out into the worksite to talk with workers about their concerns and to observe work procedures and equipment. In each workplace, the lines of responsibility from top to bottom need to be clear, and workers should know who is responsible for different health and safety issues.

V. The Importance of Training

Workers often experience work-related health problems and do not realize that the problems are related to their work, particularly when an occupational disease, for example, is in the early stages. Besides the other more obvious benefits of training, such as skills development, hazard recognition, etc., a comprehensive training programme in each workplace will help workers to:

- recognize early signs/symptoms of any potential occupational diseases before they become permanent conditions;
- assess their work environment;
- insist that management make changes before hazardous conditions can develop.

VI. Role of the Health and Safety Representative

As health and safety representative the role is to work proactively (this means taking action before hazards become a problem) to prevent workers from being exposed to occupational hazards. Management eliminates hazards or keeps them under control when they cannot be eliminated.

The hazards that may be associated with these jobs are :-

1. **Welder** - A welder can be burnt from the sparks and there is always the danger of the work process starting a fire. There is the problem of the intense light which can cause permanent eye damage as well as the fumes given off by the process which can damage the lungs.
2. **Mechanic** - Depending on the precise nature of a mechanic's duties, there may be safety problems from cuts and falls, etc., and exposure to chemical hazards: oils, solvents, asbestos and exhaust fumes. Mechanics can also have back and other musculoskeletal problems from lifting heavy parts or bending for long periods.
3. **Port worker** - Again hazards depend largely on the nature of the job and, in particular, the cargo being handled. Port workers often have no idea of

the dangerous nature of the cargo; there may be a sign on the side of a box or drum, but the information may not be in their language or in words that make much sense to the average worker. The condition of the cargo is also important as leaking drums or split bags can be very hazardous for the handlers. Other risks include falls, cuts, back and other musculoskeletal problems as well as collisions with fast moving vehicles such as fork-lift trucks or delivery trucks.

4. **Textile worker -** The textile worker faces a variety of problems. First there is the problem of safety with many machines around that are often unguarded, as well as the risk of fire with so much combustible material in the workplace. Then there are the hazards of noise and vibration. There is also exposure to dust from the material which can seriously affect the lungs. Exposure to cotton dust can lead to the occupational disease known as byssinosis.
5. **Tractor driver -** One of the most serious problems with tractors is that they often overturn and, if they have no safety cab, the driver can easily be crushed. Other problems include noise, vibration and exposure to chemical herbicides and pesticides when being sprayed by tractor.
6. **Agricultural worker -** When spraying crops, the worker may be exposed to hazardous chemicals contained in the spray. Many pesticides and herbicides that have been banned in some countries because of their toxic effects are still used in many developing countries. If spraying takes place on a windy day, the spray can be breathed into the lungs and blown on to the skin where it can cause damage. It can also be absorbed into the body through the skin.
7. **Electronics assembly worker -** An electronics assembly worker can suffer eye problems from doing close work, often in poor light. Because such workers sit still for long periods with inadequate seating, they can also suffer from back and other musculoskeletal problems. For some workers there are the dangers of solder fumes or solder "flecks" in the eye when the excess solder is cut off with pliers.
8. **Office worker -** Many people may think that office workers have no health and safety problems; this is far from true. Stress is one of the most common complaints, as well as exposure to chemical hazards from office machines such as photocopiers. Poor lighting, noise and poorly designed chairs and stools can also present problems.
9. **Construction worker -** Construction workers face a variety of hazards, particularly safety problems such as falls, slips, trips, cuts, and being hit by falling objects. There are also dangers from working high up, often without adequate safety equipment, musculoskeletal problems from lifting heavy objects, as well as the hazards associated with exposure to noisy machinery.

10. **Miner -** The hazards of mining are well known and include the ever present danger of dusts, fire, explosion and electrocution, as well as the hazards associated with vibration, extreme temperatures, noise, slips, falls, cuts, etc.

Although risks vary by type of occupation, there are several that most health care workers have in common:

- **Physical hazards.** This includes lifting or moving patients; moving or handling supplies and equipment; working in conditions of heat or cold; with hazardous machinery or tools; or areas where there are high noise levels.
- **Sharps/blood and body fluids.** Needles, syringes and many instruments used in patient care are common sources of injury and infection.
- **Radiation and medicines.** Chemotherapy, antibiotics, and radioactivity may have harmful side-effects from long-term use.
- **Chemical hazards.** The list of chemicals present in any hospital or medical centre is lengthy. They include laboratory reagents, cleaning and disinfecting agents, paints, coatings, solvents, caustics, and pesticides.
- **Infectious hazards.** Patients and even other staff or laboratories can expose staff to infectious diseases.
- **Latex hazards.** Natural rubber latex allergies can result in different types of symptoms, such as hand dermatitis or blisters, sneezing, watery eyes, asthma, or in rare cases, death.

Occupational Hazards for Women

Conditions which may lead to permanent chronic ill-health in women are low social and economic status combined with poor environmental conditions which often include outdoor work. Their focus is not exclusively on women, but of a more general nature.

Articles include the following key points:

- Pesticide exposure and reproductive outcomes: The reproductive risks and outcomes from exposure to toxic substances is methodologically difficult, but essential, given the rapidly increasing use of chemicals in industry and agriculture.
- Exposure to neurotoxins in the microelectronics industry: Behavioural problems in women is traditionally attributed to mass hysteria may result from occupational exposure to neurotoxins.
- Psychological and ergonomic stressors in garment workers: Repetitive motions and fast work speed in factories or at home are increasingly linked to physical and psychological ailments. The study concludes that more pieceworkers are disabled than fixed-wage workers, and repetitive strain injury is common among assembly line workers.

- Health consequences of 'maquiladora' work: Subjective and social factors appear to be an important determinant of women's health status. This is a consideration often neglected in epidemiological studies.
- Repetitive strain injury and occupational tasks: This disease is real and affects a large number of women. Changes in the work organisation can only temporarily alleviate painful symptoms.

Therefore, work environments which provoke these, have to be identified before the disease develops.

- Silicosis: Silicosis in women, although rare in developed countries, is usually contracted in the pottery industry and tends to progress more rapidly in women than in men. Many women in developing countries are exposed to silica dust through pottery work or hand-grinding grain.
- Cassava processing and cyanide poisoning: The importance of exposure monitoring and protective measures is outlined for this agricultural process dangerous to women's health.
- Fetal protection policies: "Fetal protection" policies often serve to protect the employer rather than the employee or the fetus.

National Implementing Legislation

Different states take different approaches to legislation, regulation, and enforcement.

In the European Union, member states have enforcing authorities to ensure that the basic legal requirements relating to occupational safety and health are met. In many EU countries, there is a strong co-operation between employer and worker organizations (e.g. Unions) to ensure good OSH performance; this has benefits for both the worker (through maintenance of health) and the enterprise (through improved productivity and quality). In 1996 the European Agency for Safety and Health at Work was founded.

In the United States, the Occupational Safety and Health Act of 1979[1] created both the National Institute for Occupational Safety and Health (NIOSH) and the Occupational Safety and Health Administration (OSHA). OSHA, in the U.S. Department of Labor, is responsible for developing and enforcing workplace safety and health regulations. NIOSH, in the U.S. Department of Health and Human Services, is focused on research, information, education, and training in occupational safety and health. OSHA has been regulating occupational safety and health since 1971. Occupational safety and health regulation of a limited number of specifically defined industries was in place for several decades before that, and broad regulations by some individual states was in place for many years prior to the establishment of OSHA.

In Canada, workers are covered by provincial or federal labor codes depending on the sector in which they work. Workers covered by federal legislation (including those in mining, transportation, and federal employment) are covered by the Canada Labor Code; all other workers are covered by the health and safety legislation of the province they work in. The Canadian Centre for Occupational Health and Safety (CCOHS), an agency of the Government of Canada, was created in 1978 by an Act of Parliament. The Act was based on the belief that all Canadians had "...a fundamental right to a healthy and safe working environment." CCOHS is mandated to promote safe and healthy workplaces to help prevent work-related injuries and illnesses.

In Malaysia, the Department of Occupational Safety and Health (DOSH) under the Ministry of Human Resources is responsible to ensure that the safety, health and welfare of workers in both the public and private sector is upheld. DOSH is responsible to enforce the Factory and Machinery Act 1969 and the Occupational Safety and Health Act 1994.

Chapter **5**

Occupational Diseases

An occupational disease is any chronic ailment that occurs as a result of work or occupational activity. An occupational disease is typically identified when it is shown that it is more prevalent in a given body of workers than in the general population, or in other worker populations. Occupational hazards that are of a traumatic nature (such as falls by roofers) are not considered to be occupational diseases. Under the law of workers' compensation in many jurisdictions, there is a presumption that specific diseases are caused by the worker being in the work environment and the burden is on the employer or insurer to show that the disease came about from another cause.

Some well-known occupational diseases include:

- Asbestosis among asbestos miners and those who work with friable asbestos insulation.
- Black lung (Coalworker's pneumoconiosis) among coal miners.
- Byssinosis among workers in parts of the cotton textile industry.
- Carpal tunnel syndrome among persons who work in the poultry industry and who do data entry.
- Lead poisoning affecting workers in many industries that processed or employed lead or lead compounds. Known under various names from ancient times, these diseases were first documented as specifically due to lead by Samuel Stockhausen in 1656.
- Phossy jaw among the London matchgirls.
- Radiation sickness among persons who work in the nuclear industry.
- Radium jaw among the Radium Girls.
- Squamous cell carcinoma of the skin of the scrotum among chimney sweeps.

An occupational disease is any one of many conditions associated with the workplace. They range from pneumoconiosis in miners to repetitive strain injury in typists. Specialists in occupational medicine, besides looking after workers, are concerned with improving health and safety at work, through public health measures, industrial legislation, and so on. The first recorded observation of an occupational disease - severe colic in a lead miner - appears in the writings of Hippocrates. The Italian physician Bernardino Ramazzini, author of Diseases of Workers published in 1700, is regarded as the father of occupational medicine. He described no less than 54 dirty or dangerous trades, including Venetian mirror-making which was associated with mercury poisoning. Britain was the first country to introduce statutory notification of occupational diseases in 1895, in the late 1990s in Britain, stress was finally recognised as an occupational disease.

Silicosis

Signs and Symptoms

Silicosis is a lung disease caused by the inhalation of crystalline silica dust. Silica is a common mineral found naturally in sand and rock, such as granite and sandstone. Silica exposure is common in mines and quarries and in a number of other occupations such as construction (especially sandblasting), foundry-work, ceramics and glass-making. Silica scars lung and affects its ability to function normally. Inhalation of crystalline silica particles may also lead to other conditions, particularly tuberculosis. There is also a link between silicosis and lung cancer.

There are different types of silicosis depending on how quickly symptoms develop after exposure. The most common form is chronic silicosis, in which symptoms occur after ten years or more of overexposure. The most common symptoms are chronic dry cough and shortness of breath with physical activity.

Diganosis

In making a diagnosis of silicosis, the doctor will start by taking a detailed medical and occupational history and conducting a thorough medical examination.

The following tests may also be conducted to make a definite diagnosis:

Chest X-ray— A chest X-ray may be taken to look for any abnormalities in your lungs. In chronic silicosis, small round opaque areas appear on the lungs.

Pulmonary Function Testing (PFT) — This test involves a series of breathing maneuvers that measure the airflow and volume of air in your lungs, which allows your doctor to objectively assess the function of your lungs.

Bronchoscopy - This test involves the passage of a flexible fiberoptic scope about the diameter of a pencil into the lungs to obtain fluid and sometimes tissue samples to aid in diagnosis.

Treatment

Unfortunately, there is no cure for silicosis. However, there are a variety of treatments available to help manage symptoms and prevent complications associated with the disease.

Lung Transplant — In severe cases, lung transplant may be an option for some patients with silicosis.

Occupational Lung Disease » Byssinosis and Related Diseases

It is not only inorganic minerals and dusts that may affect the lung. The dust produced in the processing of raw cotton, flax, or hemp may cause chronic obstructive lung disease. However, this does not have a characteristic pathology, and it does not give rise to emphysema. It is unclear whether the dust from the fibres alone or the combination of cigarette smoke and fibre dust is particularly dangerous. The disease that results is known as byssinosis, or "brown lung." Workers in cotton plants in England used to complain of "Monday morning fever" and were found to suffer an easily measurable decrement in ventilatory function when they returned to work after spending a weekend away from the plant. The active particle or contaminant in the cotton dust that is responsible for the syndrome appears to be an endotoxin produced by bacteria in the fibres of cotton.

The dust from western red cedar may cause occupational asthma, and dust from the redwood and other trees may cause an acute hypersensitivity pneumonitis. Workers in the sugarcane industry may be affected by a similar syndrome, known as bagassosis; sisal workers also develop airflow obstruction.

Occupational Lung Diseases

Asbestosis

Asbestosis is a chronic progressive disease that requires high exposures to asbestos over prolonged periods of time and is characterised pathologically by interstitial fibrosis and asbestos bodies. Asbestosis is a consequence of prolonged exposure to large quantities of asbestos, a material once widely used in construction, insulation, and manufacturing. Asbestosis can cause severe shortness of breath, fatigue, chest pain, and lead to heart problems.

Berylliosis

Berylliosis (or beryllium disease) is caused by the inhalation of beryllium particles, dust or fumes. Its symptoms include coughing, shortness of breath, fatigue, weight loss or loss of appetite, fever and sweating. Medical tests may reveal abnormal lung sounds, lung scars, decreased pulmonary function, granulomas (a nodular form of chronic inflammation) and an allergy to beryllium.

Black Lung Disease

Black lung disease is the common name for coal workers' pneumoconiosis (CWP) or anthracosis, a lung disease of older workers in the coal industry, caused by inhalation, over many years, of small amounts of coal dust. Black lung disease has gone by many names, including anthracosis, black lung, black spittle, coal worker's pneumoconiosis, miner's asthma, and silicosis.

Byssinosis

Byssinosis is an occupational disease of the lungs caused by inhalation of cotton dust or dusts from other vegetable fibers such as flax, hemp, or sisal. Byssinosis is a chronic, asthma-like narrowing of the airways. Also called brown lung disease, byssinosis results from inhaling particles of cotton, flax, hemp, or jute. Byssinosis occurs almost exclusively in people who work with unprocessed cotton.

Occupational Asthma

Occupational asthma is a form of lung disease in which the breathing passages shrink, swell, or become inflamed or congested as a result of exposure to irritants in the workplace. Occupational asthma is a lung disorder characterized by attacks of breathing difficulty, wheezing, prolonged exhalation, and cough, which is caused by various agents found in the work place.

Hypersensitivity Pneumonitis

Hypersensitivity pneumonitis is an inflammation of the lung (usually of the very small airways) caused by the body's immune reaction to small air-borne particles. These particles can be bacteria, mould, fungi, or even inorganic. Hypersensitivity pneumonitis is usually an occupational disease in which exposure to organic dusts, fungus, or moulds leads to acute and over time, chronic lung disease.

Silicosis

Silicosis is a disabling, non-reversible and sometimes fatal lung disease caused by over-exposure to respirable crystalline silica. There is no specific treatment for silicosis. Removal of the source of silica exposure is important to prevent further worsening of the disease. Supportive treatment includes cough suppression medications, bronchodilators, and oxygen if needed.

An occupational lung disease is a lung condition that develops as a result of a person inhaling harmful substances at his or her place of work. Breathing is a necessary body function. However, it can also expose the airways and lungs to dusts, chemicals, vapours, smoke, gases, fumes, and mists. Inhaling certain substances can damage lungs and promote certain conditions, including asthma, asbestosis, lung cancer, emphysema, and others. All of these diseases are due to harmful particles, vapours, gases, or mists that cause damage to the lungs when inhaled during working hours for a long period of time. The American Lung Association divides occupational lung diseases into two major groups: pneumoconiosis, caused by dust that gets into the lungs, and hypersensitivity diseases, such as asthma, that are caused by the lungs' overreaction to air borne pollutants. Additionally, some cases of lung cancer and bronchitis are classified as occupational lung diseases, as is the malady called byssinosis.

Occupational lung diseases are caused by harmful particles, mists, vapours or gases inhaled while a person works. Particles in the air may cause lung problems. Often called particulate matter (PM), particles can consist of a combination of dust, pollens, moulds, dirt, soil, ashes, and soot. Particulate matter in the air comes from many sources, such as factories, smokestacks, exhaust, fires, mining, construction, and agriculture. The finer the particles are, the more damage they can do to the lungs, because they are easily inhaled deep into the lungs, where they are absorbed into the body. "Inorganic" refers to any substances that do not contain carbon, excluding certain simple carbon oxides, such as carbon monoxide and carbon dioxide. "Organic" refers to any substances that do contain carbon, excluding simple carbon oxides, sulfides, and metal carbonates. The body has several ways of getting rid of inhaled particles. In the airways, mucus coats particles so they can be coughed up more easily. In the lungs, special scavenger cells engulf most particles and render them harmless. Sometimes, the body's defenses cannot fight-off harmful inhaled particles, and disease develops.

Repeated and long-term exposure to certain irritants on the job can lead to an array of lung diseases that may have lasting effects, even after exposure ceases. Certain occupations, because of the nature of their location, work, and environment, are more at risk for occupational lung diseases than others. Contrary to a popular misconception, coal miners are not the only ones at risk for occupational lung diseases. For instance, working in a car garage or textile factory can expose a person to hazardous chemicals, dusts, and fibres that may lead to a lifetime of lung problems if not properly diagnosed and treated.

Bronchoconstriction may result from an antigen-antibody reaction, e.g., in some forms of occupational asthma; from pharmacologic mechanisms (in byssinosis), in which the deposition of particles causes the mast cells of the airways to produce bronchoconstrictors, such as histamine and slow-reacting substance of anaphylaxis (leukotrienes C4, D4, and E4); or from irritation as a reflex mechanism (e.g., in response to sulfides). Bronchitis or mucous gland hypertrophy may be induced by long-continued deposition of particles, which may lead to a minor chronic airflow obstruction. Lung cancer may result from deposition of asbestos fibres or dusts with adsorbed radom daughters. Deposition of particles in the nose may lead to rhinitis, hay fever (which may be regarded as occupationally related in an agricultural worker), septal perforation in chrome workers, and nasal cancer in furniture workers.

Health Effects of Dusts, Gases and Vapours

Nuisance dusts are relatively inert and by definition, cause no serious health effects, although they may be irritant to the upper airways. Examples include chalk, limestone, and titanium dioxide (provided they are free of toxic impurities). They may cause radiographic changes without disease. Dusts should be considered as nuisance dusts only when there is a good evidence that they are inert and free from significant health effects not when evidence for an effect is lacking. Moreover, there is now good evidence that ultrafine particles of dusts previously considered inert, such as titanium dioxide can be toxic.

Farmers Lung Disease

Farmer's lung is a type of hypersensitivity pneumonitis. Hypersensitivity pneumonitis, also known as extrinsic allergic alveolitis, is an immunologically mediated inflammatory disease of the lung involving the terminal airways. The condition is associated with intense or repeated exposure to inhaled biologic dusts.

The classic presentation of farmer's lung results from inhalational exposure to thermophilic *Actinomycets* species and occasionally from exposure to various *Aspergillus* species.

The effect of these antigens in farmers was described as early as 1713. In Britain in 1932, Campbell described a disorder of the lung caused by inhalation of dust from moldy hay. In 1964, Ramazzini and Wright described workers getting "diseases of the chest."

Thermophilic actinomycetes species include *Saccharopolyspora rectivirgula* (formerly *Micropolyspora faeni*), *Thermoactinomyces vulgaris, Thermoactinomyces viridis,* and *Thermoactinomyces sacchari,* among others. These organisms flourish in areas of high humidity and prefer temperatures of 40-60°C.

The thermophilic actinomycetes are ubiquitous organisms usually found in contaminated ventilation systems and in decaying compost, hay, and sugarcane (bagasse). Exposure to large quantities of contaminated hay is the most common source of inhalational exposure for farmers who develop farmer's lung; therefore, grain farmers are not at risk for the development of the disease. Farmer's lung is often a disease of dairy farmers who handle contaminated hay during the winter months. Most cases of farmer's lung occur in cold, damp climates in late winter and early spring when farmers use stored hay to feed their livestock.

Exposure to the causative antigens depends on the type of farming, industry, and climate in the area. Note that farming practices are changing with time and that new antigens may be introduced or disappear from a region (e.g., the disappearance of bagassosis in Louisiana sugarcane workers, the appearance of *Pseudomonas fluorescens* in machine operator's lung). The dynamic nature of this disease and the changing environment may lead to new challenges for the clinician.

In addition to the inhalational exposure to the organic dusts responsible for the hypersensitivity reaction in farmer's lung disease, allergens, chemicals, toxic gases, and infectious agents must also be considered as potential triggers of airway symptoms in symptomatic farmers. Farming is currently ranked as one of the top three most hazardous occupations, along with construction and mining.

Pathophysiology

The pathogenesis of farmer's lung depends on the intensity, frequency, and duration of exposure and on host response to the causative antigen. Both humoral and cell-mediated immune responses seem to play a role in pathogenesis. During acute episodes, acute neutrophilic infiltration is followed by lymphocytic infiltration of the airways. Levels of interleukins 1 and 8 and tumor necrosis factor-alpha are increased. These cytokines have proinflammatory and chemotactic properties. They cause the recruitment of additional inflammatory mediators, resulting in direct cellular damage and changes in the complement pathway, which provide the necessary stimuli to increase vascular permeability and migration of leukocytes to the lung.

If the acute exposure is large, a dramatic increase in inflammation leads to increased vascular permeability, which can alter the alveolar capillary units, thus promoting hypoxemia and decreased lung compliance. If the exposure is prolonged and continuous, collagen deposition and destruction of the lung parenchyma occur with resultant decreased lung volumes.

Strong evidence suggests the involvement of immune complex–induced tissue injury (type III hypersensitivity). The presence of antigen-specific immunoglobulin and complement activation and deposition in the lung also supports immune-complex or type III hypersensitivity in the pathogenesis of farmer's lung.

Cell-mediated, delayed-type hypersensitivity (type IV hypersensitivity) also plays a major role in the pathogenesis of this syndrome. The presence of lymphocytes, macrophages, and granulomas in the alveolar spaces and the interstitium supports this conclusion.

Mortality/Morbidity

The mortality rate from farmer's lung is reportedly 0-20%.

- Death usually occurs 5 years after diagnosis.
- Several factors have been shown to increase mortality rates in farmer's lung, including clinical symptoms occurring more than 1 year before diagnosis, symptomatic recurrence, and pulmonary fibrosis at the time of diagnosis.
- Comorbid factors: Although a history of smoking appears to decrease the overall risk for the development of hypersensitivity pneumonitis, a smoking history is the strongest predictor of increased respiratory symptoms once the diagnosis is made. Pre-existing bronchial hyperreactivity with airway obstruction is also a factor.

Causes

- Thermophilic actinomycetes
  - § *S rectivirgula* (formerly *M faeni*)
  - § *T vulgaris*
  - § *T viridis*
  - § *T sacchari*
  - § *Aspergillus* species.

Chapter 6

Environmental Carcinogenesis

Industrial workers have long served as sentinels for the general population with regard to environmental hazards. Although many chemicals found in the industrial setting can also be found in the environment, industrial workers often have more intense and prolonged exposures to chemicals than does the general population. Consequently, cancers in humans caused by these substances are often first noted in the workplace.

Many of the well-established and suspected chemical carcinogens were identified through occupational studies. One of the earliest examples occurred more than 200 years ago when, in 1775, Percivall Pott, a London surgeon, described a high frequency of cancer of the scrotum among chimney sweeps, a disease known at the time as "soot-wart." A century later, other scientists noted similar cancers among gas plant workers in Germany and among oil shale workers in Scotland. Some 40 years later, certain constituents of tar, soot, and oils, known as polycyclic aromatic hydrocarbons, were found to cause cancer in laboratory animals, thus identifying the specific substances causing cancer among workers in these occupations. Preventive action was taken in Denmark, where the chimney sweeps' guild, spurred by Pott's report, urged its members to take daily baths. The success of this action was noted in a report in the 1892 British Medical Journal, "Why Foreign Sweeps Do Not Suffer From Scrotal Cancer," which pointed out that the sweeps of Northern Europe seemed to benefit from this hygiene measure, but English sweeps, who apparently ignored such recommendations, continued to develop cancer.

This brief detective story became the model for many later investigations of workplace carcinogens, including observation of unusual cancers, or a high incidence of common cancers, among groups of workers; searches for responsible agents; demonstrations that the agent can cause cancer in laboratory animals; and, finally, implementation of preventive programs.

Studies of occupational groups remain an important component of the current effort to identify the causes of human cancer. An evidence regarding occupational factors that may cause human cancer can be found in a series of critical monographs published by the International Agency for Research on Cancer (IARC). Since 1971, the IARC, an agency of the World Health Organization head-quartered in Lyon, France, has published more than 60 volumes dealing with cancer risks from individual chemicals, and mixtures of chemicals, in selected occupations or industries. Individual exposures are re-considered whenever indicated by new information.

Table 1 : Chemicals, Groups of Chemicals, or Industrial Processes Carcinogenic to Humans

GROUP 1: **Industrial processes and occupational exposures that are carcinogenic to humans:**
Aluminum production Auramine manufacture Boot and shoe manufacture and repair (certain occupations) Coal gasification Coal-tar pitches Coke production Furniture manufacture Iron and steel founding Isopropyl alcohol manufacture (strong-acid process) Manufacture of magenta Nickel refining Rubber industry (certain occupations) Underground hematite mining (with exposure to radon)
Chemicals and groups of chemicals that are carcinogenic to humans:
4-Aminobiphenyl Arsenic and arsenic compounds Asbestos Benzene Benzidine N,N-bis(2-chloroethyl)-2-naphthylamine bis(chloromethyl)ether and chloromethyl ethyl ether Chromium and certain chromium compounds Mineral oils Mustard gas 2-Naphthylamine Nickel and nickel compounds Soots, tars, and oils Vinyl chloride

Table 2 : Chemicals, Groups of Chemicals, or Industrial Processes Probably or Possibly Carcinogenic to Humans

GROUP 2A: **Probably carcinogenic to humans:**
Acrylonitrile
Benzidine-based dyes
Benzo(a)pyrene
Beryllium and beryllium compounds
Cadmium and cadmium compounds
Dimethyl sulphate
Ethylene dibromide
Ethylene oxide
Formaldehyde
Ortho-Toluidine
Polychlorinated biphenyls
Propylene oxide
Silica
Styrene oxide
Group 2B: **Possibly carcinogenic to humans:**
Acetaldehyde
Acrylamide
Amitrole
Auramine (technical grade)
1,3 Butadiene
Carbon black extracts
Carbon tetrachloride
Chlorophenols
Chlorophenoxy herbicides
DDT
3,3'-Dichlorobenzidine
3,3'-Dimethoxybenzidine (Ortho-Dianisidine)
2,4-Diaminotoluene
1,2 Dichloroethane
Dichloromethane
1,4 Dioxane
Epichlorohydrin
Ethyl acrylate
Ethylene thiourea
Hexachlorobenzene
Hydrazine
Mirex
2-Nitropropane
Organolead
Polybrominated biphenyls
Styrene
Tetrachlorodibenzo-para-dioxin (TCDD)
1,1,2,2 Tetrachloroethane
Tetrachloroethylene
Toluene diisocyanates
Toluidine
Toxaphene
Urethane

Table 3 : Cancers Associated with Various Occupations or Occupational Exposures

CANCER	SUBSTANCES OR PROCESSES
Lung	Arsenic, asbestos, bis(chloromethyl)ether, chromium compounds, coal gasification, mustard gas, nickel refining, foundry substances, radon, soots, tars, oils, acrylonitrile, beryllium, silica
Bladder	Aluminum production, auramine and magenta manufacture, rubber industry, leather industry, 4-aminobiphenyl, benzidine, naphthylamine
Nasal cavity and sinuses	Formaldehyde, isopropyl alcohol manufacture, mustard, gas, nickel refining, leather dust, wood dust
Larynx	Asbestos, isopropyl alcohol, mustard gas
Pharynx	Formaldehyde, mustard gas
Mesothelioma	Asbestos
Lymphatic and hematopoietic system	Benzene, ethylene oxide, chlorophenols, chlorophenoxy, herbicides, X-radiation
Skin	Arsenic, coal tars, mineral oils
Soft-tissue sarcoma	Chlorophenols, chlorophenoxy herbicides
Liver	Arsenic, vinyl chloride

The evaluations in the IARC monographs are based on epidemiologic data from studies in humans, bioassays in animals, and data from short-term tests and laboratory experiments. When there is enough evidence from epidemiologic studies to support a causal association with cancer, the chemical, chemical group, process, or exposure is assigned to Group 1 (e.g., asbestos, benzene, chromium, vinyl chloride, coke production, furniture manufacturing, and nickel refining). Chemicals that are carcinogenic in laboratory animals, but for which human data may be limited or lacking, are typically placed in Group 2. Those considered to be probably carcinogenic to humans appear in Group 2A (e.g., acrylonitrile, cadmium, formaldehyde, and silica), and those considered to be possibly carcinogenic to humans are placed in Group 2B (including butadiene, carbon tetrachloride, chlorophenoxy herbicides, DDT, styrene, and tetrachloroethylene). The other categories are considered not classifiable as to carcinogenicity (Group 3), or not carcinogenic to humans (Group 4).

The agents carcinogenic to humans (those listed in Table 1) and those probably or possibly carcinogenic to humans (Table 2) do not fall into any particular class of substances. They include metals (arsenic, chromium, nickel, cadmium), solvents (benzene, styrene, carbon tetrachloride, dichloromethane), organic and inorganic dusts (leather or wood dusts, asbestos, silica), chemicals used to construct polymers (acrylonitrile, formaldehyde, vinyl chloride), and pesticides (ethylene oxide, amitrole, chlorophenoxy herbicides, DDT, toxaphene). Tumors in humans associated with occupational exposures from substances or processes listed Tables 1 and 2 are shown in Table 3. Efforts to identify and clarify cancer risks associated with the workplace continue. This research is necessary to ensure a safe work environment and to identify environmental factors that may cause cancer in the general population.

The Fraction of Cancer Attributable to Occupation

In developed countries, 30% of cancers are attributed to diet or nutrition, 16% to tobacco, 8% to infections, 5% to occupational exposures, 2% to environmental pollution, and 39% to other causes. "Of course, the major exposure to carcinogens is not through variations in the external environment (e.g., air, water, radiation) so much as in differences in lifestyle (e.g., reproduction, diet, tobacco use)."

Causal Links to Occupation for Specific Cancers

The following cancers have been associated with occupational exposures.

Bladder Cancer

A sentinel health event (occupational) associated with rubber and dye workers who were exposed to benzidine, alpha- and beta-naphthylamine, magenta, auramine, 4-aminobiphenyl, and 4-nitrophenyl; [Mullan] Beta-naphthylamine, benzidine, 4-aminobiphenyl were discontinued from industrial production and use. Suspected bladder carcinogens still in use include benzidine-based dyes, o-toluidine, 4-chloro-o-toluidine, and MBOCA. Other occupational exposures associated with bladder cancer include PAHs (aluminum production, coal gasification, coal-tar pitches, benzo[a]pyrene, and diesel exhaust), and hair dyes (hairdresser or barber). Diesel exhaust had a positive association in many case-control studies, but was negative in cohort studies of transportation workers. "An open question is whether occupational exposures in industries identified in the past as high risk can still be linked to an excess risk of bladder cancer." Occupational carcinogens with strong evidence include 4-aminobiphenyl, benzidine, and 2-naphthylamine, aluminum production, coal gasification, magenta manufacture, and the rubber industry. Causes for Bladder Cancer are as follows:

1. Benzidine
2. 4-Aminodiphenyl
3. Beta-Naphthylamine
4. Aluminum production
5. Coal gasification
6. Magenta manufacture
7. Rubber industry

Bone cancer

A sentinel health event (occupational) associated with radium exposure by radium chemists, processors, and dial painters; Documented causes of radiation-induced bone cancer in humans include 226,228-radium, 224-radium, plutonium, Thorotrast (thorium oxide), and radiotherapy. "Radiation-induced bone cancer appears, it seems, only at very high doses, and it is rarely reported at doses under 5 Gy." There was a strong evidence of work-related bone cancer in radium dial painters and a "meaningful association" in Mayak workers who had very high exposure to plutonium. Bone cancer is due to -

1. Ionizing radiation
2. Radium
3. Plutonium
4. Brain cancer

Ionizing radiation, at high doses well above background levels, is the only established environmental cause of brain cancer. Some epidemiological studies have found an increased risk of brain cancer associated with petrochemical workers, rubber and tire workers, electrical workers, health professionals, farmers, and workers with exposures to vinyl chloride and polychlorinated biphenyls. Brain cancer after radiation treatment is "occasionally associated with robust risk estimates." It is seen mainly after high-dose irradiation in childhood, e.g., an estimated dose of 1.5 Gy for treatment of ringworm of the scalp. Brain cancer was not associated with radiation in studies of Japanese A-bomb survivors, radium dial painters, radiologists, underground miners, nuclear workers, and uranium processors. There is a suggestive evidence that non-arsenical insecticides and petroleum refining are occupational carcinogens. There is no strong evidence that occupational exposure to ionizing radiation is associated with brain cancer.

Colorectal Cancer

Studies showing an association with asbestos exposure are inconsistent. Increased colorectal cancer has been reported in workers exposed to mineral oils in the operation of printing, typesetting, and textile processing machines. There was an association between rectal cancer and machinists exposed to mineral-based metalworking fluids. Colon cancer is increased in white-collar workers who have sedentary jobs. Some studies have shown a small excess risk for workers exposed to asbestos. Colon cancer was not associated with ionizing radiation exposure in studies of radiologists, underground miners, nuclear workers, or uranium processors. There was a "meaningful association" in Japanese A-bomb survivors.

Oesophagus Cancer

Information relating to occupational causes of oesophageal cancer is limited. It is not known what proportion of oesophageal cancer today is linked to workplace exposures. Some studies suggest associations with occupational exposures to perchloroethylene, mustard gas, silica dust, metal dust, asbestos, combustion products, sulfuric acid, carbon black, and ionizing radiation. Many of these studies have inadequate adjustment for the main confounders, smoking and alcohol use. Based on heavier exposure levels that existed in the past, two groups that showed higher risks for cancer of the oesophagus were workers who vulcanized rubber and those who assembled automobiles. There is a suggestive evidence that occupational exposure to soots and tetrachloroethylene cause oesophageal cancer. Studies of radiologists, underground miners, and nuclear workers have found negative associations. "Meaningful associations" were found for Japanese A-bomb survivors and patients with spondylitis treated with medical radiation.

Liver Cancer

There were negative associations that ionizing radiation caused liver cancer in studies of radium dial painters, radiologists, underground miners, nuclear workers, uranium processors, and Mayak workers. There was a strong evidence that Thorotrast (Th-232) caused liver cancer when used as a radiographic contrast agent between 1928 and 1955. Hepatocellular carcinomas (HCC) comprise about 75% to 90% of liver cancers. "Eighty to ninety five percent of HCCs are associated with chronic infection of hepatocytes with either of two viruses, hepatitis B virus (HBV) or hepatitis C virus (HCV)." Increased risk of angiosarcoma of the liver (ASL) was found in workers exposed to vinyl chloride while cleaning reactor vessels for the production of polyvinyl chloride. There is a suggestive evidence that German vineyard workers exposed to arsenic in the 1930s and 1940s had increased incidence of angiosarcoma of the liver. Cause of liver cancer is Vinyl chloride

Kidney Cancer

"Kidney cancer, in contrast to bladder cancer, is generally not considered an occupation-related cancer." The relationship between occupational exposures and kidney cancer is poorly understood. Increased risks have been reported in the following industries: coke production, iron and steel production, chimney sweeping, nickel smelting, lead smelting, dry cleaning, oil refining, textile production, electric power, farming and printing. Also in the following jobs: tailors, firefighters, painters, gasoline station attendants, truck drivers, and workers exposed to asbestos.

Laryngeal Cancer

Cancer of the larynx in western countries is for the most part attributable to smoking tobacco and drinking alcohol. Possible occupational risks include exposure to nickel, asbestos, and ionizing radiation. Studies of cohorts from past decades have found strong associations between laryngeal cancer and occupational exposures to mustard gas manufacturing, nickel refining, and metalworking with mineral-based oils. Risk factors for laryngeal cancer include smoking, drinking alcohol, and asbestos exposure. Possible risk factors are leather workers, textile workers, and exposure to gasoline, diesel oil, and mineral oil. There is a strong evidence that the following are occupational carcinogens: isopropanol manufacture, strong acid process; inorganic acid mists containing sulfuric acid; and mustard gas. There is a suggestive evidence that exposure to asbestos and the rubber industry are associated with an increased risk of work-related laryngeal cancer. "The committee concluded that the evidence is sufficient to infer a causal relationship between asbestos exposure and laryngeal cancer." Cause of laryngeal cancer is due to

1. Sulfuric acid mists
2. Asbestos
3. Mustard gas

Leukemia

Leukemias represent 3% of all malignant neoplasms. Ionizing radiation, benzene, and cytotoxic drugs are known causes of acute leukemia. In atomic bomb survivors, the incidence of acute leukemias peaked at 2 to 5 years and declined after 10 years. After chemotherapy, the incidence peaks at 5 to 8 years. Acute lymphocytic leukemia (ALL) is limited mainly to the first 2 decades of life. It is associated with several congenital disorders characterized by chromosomal abnormalities. Since a majority of patients with these congenital syndromes do not develop ALL, an environmental cause is suspected. There is a strong evidence for associations between the following and occupational leukemia: boot and shoe manufacture and repair, benzene, ionizing radiation, and ethylene oxide. In studies of ionizing radiation and leukemia, strong associations were found for Japanese A-bomb survivors, radiation treatments for spondylitis, and use of Thorotrast as a contrast agent. Meaningful associations were found for radiologists working in past decades and Mayak workers. Negative associations were found for radium dial painters, underground miners, uranium processors, and Chernobyl cleanup workers. For nuclear workers the association is "suggested but unconfirmed or questionable association." Ionizing radiation increases risks for acute lymphocytic and acute myeloid leukemias and chronic myeloid leukemia, but not for chronic lymphocytic leukemia. Leukemia is due to following reasons:

1. Ionizing radiation
2. Benzene
3. Ethylene oxide

Lung cancer

Processes strongly associated with occupational lung cancer: aluminum production, coke production, coal gasification, underground hematite mining (radon), iron and steel founding, nickel refining (nickel oxides and sulfides), painters, and passive smoking. Agents (IARC Group 1) strongly associated with occupational lung cancer: arsenic compounds, hexavalent chromium compounds, asbestos, beryllium, cadmium compounds, ionizing radiation, crystalline silica, soots, and talc containing asbestiform fibers. Bis(chloromethyl)ether and chloromethyl methyl ether (technical grade) were strongly associated with lung (oat cell) cancer. Ionizing radiation was strongly associated with lung cancer in studies of Japanese A-bomb survivors, patients treated with radiation for Hodgkin's disease, underground miners, and Mayak workers with heavy exposure to plutonium. Negative associations were found in studies of radiologists, nuclear workers, uranium workers, and populations exposed to background radiation. Lung cancer is caused due to -

1. Aluminum production
2. Arsenic and compounds
3. Asbestos
4. Beryllium and compounds
5. Cadmium and compounds
6. bis-Chloromethyl ether
7. Chloromethyl methyl ether
8. Chromium compounds, hexavalent
9. Coal gasification
10. Coke oven emissions
11. Hematite mining, underground, with radon exposure
12. Involuntary (passive) smoking
13. Ionizing radiation
14. Iron and steel founding
15. Selected nickel compounds, including combinations of nickel oxides and sulfides in the nickel refining industry;
    (a) Painters
    (b) Plutonium
    (c) Radon
    (d) Silica, crystalline

(e) Soots
(f) Talc containing asbestiform fibers
(g) Melanoma
(h) Solar radiation
(i) Mesothelioma, peritoneal
(j) A sentinel health event (occupational) associated with asbestos exposure; [Mullan]
(k) Asbestos
(l) Erionite
(m) Mesothelioma, pleural

Seventy percent of patients with pleural mesothelioma reported in Denmark around 1980 were excess cases due to occupational asbestos exposure assuming a baseline incidence rate for women in rural areas. This is due to-

1. Asbestos
2. Erionite
3. Nasal sinus cancer

Agents associated with sino-nasal cancer include cigarette smoking, wood and leather dust, nickel refining, chromates, mustard gas manufacturing, isopropanol manufacturing (sulfuric acid mists), and possibly formaldehyde and welding. Softwood dust is associated with squamous cell carcinoma, and hardwood dust is associated with adenocarcinoma of the nasal cavity. An increased risk exists for sawmill workers, furniture workers, wood products workers, and carpenters. No increased risk exists for workers in forestry, logging, or paper and pulp. Seventy percent of patients with sino-nasal adenocarcinoma reported in Denmark between 1965 and 1974 had worked for many years in wood-working jobs. Sino-nasal cancer is associated with occupational exposure to nickel (refining), wood dusts (furniture making), boot and shoe dusts (manufacturing), hexavalent chromium (pigment manufacturing), and radium (dial painting). Strong evidence: Boot and shoe manufacture and repair; furniture and cabinet making; isopropanol manufacture, strong acid process (sulfuric acid); nickel refining (nickel oxides and sulfides); wood dust. Suggestive evidence: chromium compounds, hexavalent; formaldehyde, mineral oils, untreated and mildly treated. People affected by lung cancer are :

1. Boot and shoe manufacture and repair
2. Furniture and cabinet making
3. Isopropanol manufacture, strong acid process (sulfuric acid)
4. Nickel refining (nickel oxides and sulfides)
5. Wood dusts
6. Radium (dial painting)

Nasopharynx Cancer

There is a suggestive evidence that mustard gas and formaldehyde can cause occupational nasopharyngeal cancer. Formaldehyde is a Group 1 carcinogen with sufficient evidence that it caused nasopharyngeal cancer in humans.

Pancreatic Cancer

For pancreatic cancer, occupation is not a major causal factor and probably accounts for less than 5% of all cases. There is a weak association with chlorinated hydrocarbons in epidemiological studies. Increased risk of pancreatic cancer was associated with working as a dry cleaner. There is inconclusive evidence that workers exposed to cadmium, asbestos, and ionizing radiation have increased risk. There is a suggestive evidence that acrylamide can cause pancreatic cancer.

Skin Cancer

The major risk for outdoor workers is exposure to ultraviolet light. Other agents carcinogenic to the skin include: PAHs (coal tar, shale oil, or mineral oils); arsenic (pesticide manufacturing; sheep dip; copper, lead or zinc smelting); and ionizing radiation (radiologists); Arsenic exposure is associated with an increased risk of basal cell cancer after a long latency. Sun exposure increases risk for basal cell cancer, squamous cell cancer, and melanoma. Chronic arsenic poisoning causes keratoses of palms and soles, patchy hyperpigmentation, and skin cancer (squamous and basal cell). The evidence is strong for associations between the following agents or processes and occupational skin cancer: arsenic and compounds; coal tars and pitches; coal gasification; coke production; dibenz[a,h]anthracene; mineral oils, untreated and mildly treated; shale oils or shale-derived lubricants; solar radiation; and soots. Studies of ionizing radiation and skin cancer have found "meaningful associations" for Japanese A-bomb survivors, tinea capitis patients treated with radiation, and radiologists working in earlier decades. Skin cancer is caused due to :

1. Solar radiation
2. Ionizing radiation
3. Arsenic and compounds
4. Soots
5. Coal tars and pitches
6. Mineral oils, untreated and mildly treated
7. Shale oils
8. Stomach cancer

In the list of industrial processes associated with human cancer, only working in the rubber industry is a possible cause of stomach cancer. Populations exposed to high-dose radiation from the atomic bomb and from radiotherapy for ankylosing spondylitis had increased risk for stomach cancer. "Because the large literature on occupational exposures and gastric cancer risk is not strikingly consistent, the data need cautious interpretation." For the following chemicals, the evidence is weak for a causal relationship: asbestos, silica, wood dust, chlorophenols. Many studies show a positive association between gastric cancer and occupational exposure to mineral/metal dusts, nitrosamines, and some metalworking fluids. Substantial evidence exists for a causal association between heavy exposure to asbestos and stomach cancer. Strong evidence exists for coal miners, and some evidence exists for ethylene oxide production, painters, and exposure to sulfates and sulfites in the pulp and paper industry. There was suggestive evidence of increased occupational stomach cancer in painters and workers in the rubber industry. Negative associations were found in studies of ionizing radiation and stomach cancer in radiologists, underground miners, nuclear workers, and uranium processors.

Thyroid Cancer

Thyroid cancer was strongly associated with exposure to ionizing radiation in Japanese A-bomb survivors, patients treated with radiation to the head and neck (tinea capitis, thymus, and hemangiomas); and children exposed to Chernobyl fallout. Negative associations have been found in studies of nuclear workers, uranium processors, Chernobyl cleanup workers, and populations exposed to background radiation.

General Methods of Prevention of Occupational Diseases

The main aim of preventive measure is to: Eliminate the HAZARD. If this is not possible then we try to eliminate the RISK, i.e., eliminate the possibility that the HAZARD can induce Health effects.

The body of prevention system is supported by two legs:

Environmental (engineering) + Medical measures.

I) Environmental (Engineering) Measures:

Substitution: Substitution of hazardous substances or operation by non-hazardous ones, e.g., fiberglass in place of asbestos, compression instead of hammering.

Isolation : Isolation of the hazardous process inside the work places so that other surrounding workers are not affected, e.g., isolation of radiation or noise sources

Segregation : Segregation of the hazardous process away from the work places, e.g., incinerators (sources of gases), electric generators (sources of noise), ovens (sources of heat).

Ventilation : By fans to increase air movement or by exhaust system for suction of toxic gases or dust to be collected in a special disposal system.

Environmental monitoring : To assure that exposure at the work places does not exceed the allowable levels or concentration [Threshold Limit Value (TLV)].

Assurance of Ergonomics in the work place: To adapt the work situation to the physical capabilities of the worker. This is to avoid loss of effort and time and the development of accidents or musculo skeletal disorders.

(II) Medical Measures:

Occupational oriented Pre-employment medical examination which aims at:

- To have a baseline medical data to be compared with the results of periodic medical examination.
- To advice the worker as regards the suitability of the work place to his health status.

Periodic medical examination : The type and periodicity of the examination is done according to the exposure. e.g. audiometry for noise exposure, lung X-ray and ventilatory function for silica exposure. This aims at:

- Early detection of health effects of hazardous exposure to be treated in its reversible stage.

Monitoring the efficiency of occupational health services. Occurrence of cases of occupational diseases indicate deficiency of the system of services. This is called Biological monitoring.

This is carried by measuring the toxic element or one of its metabolites or its effects in the biological fluids of the workers.

Examples include:

- Blood lead and urine delta aminoleveolonic acid in lead exposed workers.
- Urine mercury, in mercury exposed workers.
- Blood count for X-ray workers.
- Hippuric acid in urine of toluene exposed workers.
- Urine Phenol in benzene exposed workers.

Personal protective devices : To prevent the penetration of the uncontrolled hazardous exposures in the body, e.g., mask for dust or gases exposure, ear muffs or earplugs for noise exposure, masks for dust exposure , gloves for chemical exposure and leaded aprons for radiation exposure etc.

Note: Protective devices must be suitable, comfortable, do not interfere with job continuity and periodically maintained and workers should know their importance in protecting their health otherwise, it will not be used.

Health education : Workers should be informed about the hazards of their job and how to avoid the health effects of such hazards. This is called the *"Wright to know"*.

Occupational Health Laws

The Factories Act, 1948, the Mines Act, 1952, the Dock Workers (Safety, Health & Welfare) Act, 1986 are some of the laws which contain provisions regulating the health of workers in an establishment. Whereas the Employees State Insurance Act, 1948 and the Workmen's Compensation Act, 1923 are compensatory in nature.

Health Provisions under the Factories Act, 1948

The Factories Act, 1948 was enacted with the object of protecting workers from subjecting to unduly long hours of bodily strain or manual labour. It lays down that employees should work in healthy and sanitary conditions so far as the manufacturing will allow and that precautions should be taken for their safety and for the prevention of accidents.

The Act defines a 'worker' as any person employed directly or through any agency (including a contractor), whether for remuneration or not in any manufacturing process or in any work incidental to or connected with the manufacturing process. It is required that work performed should be connected with the product which is produced in the manufacturing process.

Section 10 of the Act lays down that a State Government may appoint qualified medical practitioners as 'certifying surgeons' to discharge the following duties:

a) Examination and certification of young persons and examination of persons engaged in 'hazardous occupation'.

b) Exercising medical supervision where the substances used or new manufacturing processes adopted may result in a likelihood of injury to the workers.

c) Exercising medical supervision in case of young persons to be employed in work likely to cause injury.

Twenty seven processes and operations have been identified as dangerous in The *Maharashtra Factories Rules, 1963*. These Rules lay down detailed instructions regarding preventive measures, protective devices, cautionary notices as well as medical examination of workers. The State Governments have adopted these rules depending on their local needs. The Act lists 29 occupational diseases and obliges the manager of a factory and medical practitioners to notify the Chief Inspector of Factories if any worker contacts any of the diseases. The Rules are very comprehensive in laying down special provisions with respect to health, safety and welfare of workers including medical examinations, setting up of Occupational Health Centers, etc. The only lapse has been its ineffective implementation since most of the discretionary powers lie in the hands of the

Inspectors and occupiers. Although very few cases of occupational diseases are reported in factories, the working conditions in most of the factories handling hazardous chemicals have higher risk potential.

The Employees' State Insurance (ESI) Act, 1948

It is a social security legislation enacted with the object of ameliorating various risks and contingencies sustained by workers while serving in a factory or establishment.

It is designed to provide cash benefit in the case of sickness, maternity and employment injury, payment in the form of pension to the dependents of workers who died of employment injury and medical benefit to workers. It recognizes the contributory principle against such contingencies, provides protection against sickness, replaces lumpsum payments by pension in the case of dependents benefit and places the liability for claims on a statutory organization.

The Act does not cover 'seasonal employments'. It defines 'employment injury' as personal injury to employees, caused by accident or occupational diseases, in an insurable employment.

The Act lays down provisions to set up an ESI Corporation, to promote measures to improve health and welfare of insured persons and a Medical Benefit Council to advise the Corporation on medical benefits, certification, etc. The Medical Boards have to ascertain the percentage of disability of injured workers before submitting their report to the Corporation in order to grant compensation to the workers. An injured worker has to wait for months before the Medical Board calls him for a check-up.

The main source of revenue for the ESI Fund is the Contribution paid by the employers and the employees. The purposes for which the Fund is to be used are numerous. It includes payment of benefits, provision of medical treatment to insured families, meet charges in connection with medical treatment, maintenance of hospitals, dispensaries, etc. In existing conditions there is gross misuse of these funds.

The discretionary powers with respect to using the Fund amount lie solely with the Corporation along with the State Governments. According to the Occupational Health and Safety Center, Mumbai, the Corporation has only 4 occupational disease centers for workers.

Section 39 of the Act makes the employer primarily liable for the payment of contribution on behalf of himself and his employees towards the ESI Fund. In case of misuse of the contribution by employer, the employee can sue the employer in the Employees' State Insurance Court set up by the respective State Government.

Where an employee makes a claim on the grounds of sickness, disablement or maternity, it has to be made against the ESI Corporation and not against the employer. The process involved to obtain the compensation is tedious. Such a lapse renders the very object of the Act to provide for quick claims as unreal.

Under the Workmen's Compensation Act, 1923, there exists a legal obligation on the employer to pay compensation to workmen involved in accidents arising during the course of their employment. The prerequisites for payment of compensation to such workmen are as follows:

* Personal injury must be caused.
* There must be temporary, total or partial disablement due to an accident, which also includes occupational diseases.

The State Government has to appoint a Commissioner to decide the liability of an employer to pay compensation, the amount and duration of compensation, among other issues. An appeal may lie to the High Court in case the applicant is grieved with the Commissioner's orders.

Compensation is decided on the nature of injury caused. Where the injury from an accident results in the death of the workman, the minimum compensation payable is around Rs. 50,000 and the maximum may extend to Rs. 3 lacs. In case of permanent total disablement and permanent partial disablement, compensation may extend to Rs. 60,000, depending on its nature. Further, the amount of compensation is calculated on the wage-group to which the workman belongs and the time-period for which he has worked.

There is no comprehensive law on occupational health, though the Central Government has in its various policies stressed the need to effectively implement the existing laws.

Chapter 7

Industrial Toxicology

Chemical Toxicity

Toxicology is the study of the nature and action of chemical substances on living systems. Toxicity is the ability of a chemical molecule or compound to produce injury once it reaches a susceptible site in or on the body. Hazard is the probability or likelihood that injury will occur considering the manner in which the substance is used.

Dose-Response Relationships

The potential toxicity (harmful action) inherent in a substance is manifest only when that substance comes in contact with a susceptible living biological system. A chemical normally thought of as "harmless" will evoke a toxic response if added to a biological system in sufficient amount. The toxic potency of a chemical is defined by the relationship between the dose (the amount) of the chemical and the response that is produced in a biological system.

Routes of Entry into the Body

There are three main routes by which hazardous chemicals enter the body:

- absorption through the respiratory tract through inhalation.
- absorption or injection through the skin or eyes.
- absorption through the digestive tract. This can occur through eating or smoking with contaminated hands or in contaminated work areas.

Most exposure standards, Threshold Limit Values (TLVs) and Permissible Exposure Limits (PELs), are based on the inhalation route of exposure. They are normally expressed in terms of either parts per million (ppm) or milligrams per cubic meter (mg/m^3) concentration in air.

If a significant route of exposure for a substance is through skin contact, the MSDS will have a "skin" notation associated with the listed exposure limit. Examples include: some pesticides, carbon disulfide, phenol, carbon tetrachloride, dioxane, mercury, thallium compounds, ethylene, hydrogen cyanide.

Health Effects

Acute poisoning is characterized by rapid absorption of the substance and the exposure is sudden and severe. Normally, a single large exposure is involved. Examples: carbon monoxide or cyanide poisoning.

Chronic poisoning is characterized by prolonged or repeated exposures of a duration measured in days, months or years. Symptoms may not be immediately apparent. Examples: lead or mercury poisoning or pesticide exposure.

"Local" refers to the site of action of an agent and means the action takes place at the point or area of contact. The site may be skin, mucous membranes, the respiratory tract, gastrointestinal system, eyes, etc. Absorption does not necessarily occur. Examples: some strong acids or alkalis.

"Systemic" refers to a site of action other than the point of contact and presupposes absorption has taken place. For example, an inhaled material may act on the liver. Example: arsenic affects the blood, nervous system, liver, kidneys and skin.

Cumulative poisons are characterized by materials that tend to build up in the body as a result of chronic exposure. The effects are not seen until a critical body burden is reached.

Example: heavy metals (such as Lead).

Synergistic responses: When two or more hazardous material exposures occur, the resulting effect can be greater than the effect of the individual exposures. Example: exposure to ***both*** asbestos and tobacco smoke, producing lung cancer or mesothelioma.

Other Factors Affecting Toxicity

Rate of entry and route of exposure (how fast the toxic dose is delivered and by what means) are important factors. A person's ***age*** can affect the capacity to repair tissue damage. ***Previous exposures*** can lead to tolerance or increased sensitivity.

General state of health, physical condition, and life style, can affect the toxic response. Pre-existing disease can result in increased sensitivity.

Industrial Hygiene

Industrial hygiene is defined as the recognition, evaluation, and control of workplace hazards. Its origins are based on limiting personal exposures to chemicals, and have evolved to address the control of most other workplace hazards including over-exposure to noise, heat, vibration, and repetitive motion.

Occupational exposure to chemicals is still considered one of the most widespread hazards in the industry. The use of engineering controls is the preferred method of limiting these exposures. Dilution and capture ventilation are two important methods to control occupational exposure. The design and position of hoods and vents, and amount of air infiltration can substantially change exposure conditions. Material Safety Data Sheets and other documentation provide a basis for predicting adverse effects, disposal needs, and fire and ignition concerns.

Information about several chemicals or groups of chemicals can be found in Reference Data Sheets written by Meridian Engineering and Technology staff members and consultants. Some of these data sheets also include specific topics in the fields of safety or environmental engineering.

Sampling and monitoring equipment is available to quantify exposures to contaminants, noise, radiation, and heat. Correct sampling strategy and interpretation are essential elements of an industrial hygiene survey. Confined spaces, ventilation changes, the performance of collection systems, and pressure relief valves are all items of concern for exposure evaluation.

The selection of appropriate personal protective equipment for exposure control often requires both an understanding of the limitations of the equipment, and the expected exposure parameters as determined by an industrial hygiene evaluation. Engineering controls are considered the preferred method of control, but personal protective equipment plays an important role for health, safety, and rescue, especially when using confined space entry procedures. The potential for misuse or misapplication of these devices should be evaluated.

Indusrial Hygiene and Safety Measures

The protection of worker health and safety presents complex challenges to business decision-makers. Regulatory emphasis is shifting from enforcement toward collaboration and partnership with business and industry. Workers compensation costs continue to be a significant business concern. While many reform efforts have brought these costs under control, insurance premiums rank among the highest costs to a company after wages and benefits. In addition, efforts to circumvent workers, compensation programs are leading to increased litigation. A strong health and safety program that emphasizes prevention of injuries and illness is the key to maintaining regulatory compliance, minimizing the impact of workers compensation losses, and reducing the potential for litigation to businesses as well as their key equipment and material suppliers. Worker health and safety is not just a laudable goal. Businesses have found that improvements in their health and safety programs can have a positive impact on their bottom line.

ENVIRON provides a wide range of specialized health and safety consulting services to international concerns ranging from multinational corporations to small businesses. Clients from the chemical, manufacturing, services, distribution, commercial, financial, and legal sectors, as well as national, state, and local agencies, rely on us to develop solutions that effectively address both business needs and health and safety program requirements. Our experienced and certified health and safety professionals offer clients comprehensive and integrated health, safety, and environmental capabilities, extensive technical expertise, and a long-standing reputation for excellence.

ENVIRON's Industrial Hygiene and Safety consulting services include:

Technical Services

- Industrial Hygiene Exposure Assessment
- Safety and Risk Management
- Industrial Accident Forensic Analysis

Building Sciences

- Indoor Environmental Quality and Microbial Assessment
- Materials Inspection and Management

Health and Safety Management and Regulatory Compliance

- Auditing and 3rd Party Performance Evaluations
- Program Development
- Education and Training
- Management Systems
- Toxicology and Risk Assessment
- OHS Professional Outsourcing
- Litigation Support and Expert Testimony
- People

Room for improvement in EU standards for industrial chemical risk assessment

Chemicals are an integral part of our daily lives. In the European Union, the effects of the use of industrial chemicals on humans and the environment are evaluated using the REACH system: Registration, Evaluation and Authorisation and restriction of Chemicals. This system is applied to all chemicals produced in or imported into the European Union. Chemical risk management includes all of the obligatory and voluntary measures used to limit the risks posed by chemicals. Risk managers in government and industry are assisted in this process by experts in various fields, including risk assessment.

Both the risk managers and their consultants admit that the issue is too complex and the available data are too limited to be able to have a thorough understanding of all of the risks involved in chemical risk assessment. An important aspect of the decision-making process is the consideration of whether precautionary measures should be taken.

The process can be improved through the application of a framework for risk assessment that includes extra attention for the uncertainties. The methodology can be improved by using a phased approach to the analysis of the uncertainties, starting with a simple approach and then moving on to a deeper analysis and a statistical approach as necessary.

Risk Assessment of Chemicals

The new regulation concerning the registration, evaluation, authorisation and restriction of chemicals (REACH) requires demonstration of the safe manufacture of chemicals and their safe use throughout the supply chain. While REACH is based on the precautionary principle, it includes also the aim to reduce animal testing where possible.

The goal of the project OSIRIS is to develop **integrated testing strategies (ITS)** fit for REACH that enable to significantly increase the use of non-testing information for regulatory decision-making, and thus to minimise the need for animal testing. To this end, operational procedures are developed, tested and disseminated that guide a transparent and scientifically sound evaluation of chemical substances in a risk-driven, context-specific and substance-tailored manner.

The envisaged decision theory framework includes **alternative methods** such as chemical and biological read-across, *in vitro* results, *in vivo* information on analogues, qualitative and quantitative structure-activity relationships, thresholds of toxicological concern and exposure-based waiving. OSIRIS takes into account cost-benefit analyses, and it aims to close the gap between risk and societal risk perception. It is based on the new REACH paradigm to move away from extensive standard testing to a more **intelligent, substance-tailored approach**.

The work is organised in the five interlinked Research Pillars:

- Chemical Domain,
- Biological Domain,
- Exposure,
- Integration Strategies and Tools,
- Case Studies,

with a particular focus on more complex, long-term and high-cost endpoints. Case studies will demonstrate the feasibility and effectiveness of the new ITS methodologies, and provide guidance in concrete form.

To ensure optimal uptake of the results obtained in this project, **end-users in industry and regulatory authorities** are closely involved in monitoring and in providing specific technical contributions to this project.

Risk Management of Industrial Chemicals

The tragic incident at Bhopal, India made it clear that safety reviews for identification and control of accidents involving toxic chemicals must be more systematic. They discuss how to perform and supervise safety studies for chemical, petrochemical, petroleum refining, and other facilities. They discuss all aspects of detection, prevention, and mitigation of risks associated with processing, handling, and production of hazardous chemicals. Special attention is given to hazard identification and hazard assessment techniques ranging from simple screening checklists to highly structured Hazard and Operability (HAZOP) analysis. It is shown how to calculate potential consequences of identified hazards, quantify the likelihood of these events, and combine equipment failure rate data and human reliability analysis with hazard assessment. One will also benefit from the book's rundowns of how to:

* apply expert systems and artificial intelligence in risk management.
* instill safety-oriented operating and maintenance procedures.
* train operators and emergency response personnel.
* conduct internal and external safety audits.
* perform chemical dispersion, explosion, and fire analyses.
* assess health effects from chemical releases.
* use insurance vehicles to deal with residual risk.

Risk Assessment and Risk Management for the Chemical Process Industry is an essential source on minimizing the dangers of toxic incidents and accidents. It is essential reading for safety engineers, regulatory managers, environmental engineers, and other professionals responsible for safety in chemical plants.

Sharing information about chemical assessments is a priority issue arising from the UN Conference on Environment and Development (UNCED) (June 1992, Rio de Janeiro). Better use of existing information and mechanisms for information exchange are vital parts of Agenda 21. The Organization for Economic Co-operation and Development (OECD) initiated efforts in the mid-1990s to encourage information and work sharing among member and other countries.

More recently, an OECD Task Force (OECD TF) on New Chemicals was established to develop and implement a two-year program of work that, among other things, promotes bi/multi-lateral arrangements to formalize co-operation. Some of the activities the Task Force is engaged in the development of a standard notification form to simplify reporting and facilitate data sharing and the development of a standard format for assessment of new chemicals. By the same token, the OECD TF is working towards the harmonization of data requirements for certain special categories (i.e. site-limited, export only, research and development), in order to provide greater transparency of such requirements and to reduce the resources needed to manage new industrial chemical programs without compromising the protection of the environment and human health.

This Arrangement is in keeping with wider OECD effort aimed at learning from each other, enhancing information and work sharing, and harmonizing national new industrial chemical schemes. The demonstration of international leadership in this regard will ultimately serve to leverage government resources in both countries through similar Arrangements with other countries. The benefits associated with such a bilateral Arrangement would be to increase the efficiency of new industrial chemical notification and assessment schemes by providing greater transparency in assessments. It can also lead to a possible reduction in animal testing, a reduction in resources needed for new industrial chemical works in government and industry, and speed up of product introduction for some chemicals.

This Arrangement between Australia and Canada (the Participants) and the activities undertaken pursuant there to, will provide a model for the co-operation envisaged with other OECD countries and will be consistent with the initiatives undertaken by the OECD TF on New Chemicals. Through this work, a wealth of experience will be gained regarding new industrial chemicals notification and assessment schemes between the Participants.

Threshold Limit Value

The Threshold Limit Value (TLV) of a chemical substance is a level to which it is believed a worker can be exposed day after day for a working lifetime without adverse health effects. Strictly speaking, TLV is a reserved term of the American Conference of Governmental Industrial Hygienists (ACGIH). It is, however, sometimes loosely used to refer to other similar concepts used in occupational health and toxicology. TLVs (along with biological exposure indices or BEIs) are published annually by the ACGIH.

The TLV is an estimate based on the known toxicity in humans or animals of a given chemical substance and the reliability and accuracy of the latest sampling and analytical methods. It is not a static definition since new research can often modify the risk assessment of substances and new laboratory or instrumental analysis methods can improve analytical detection limits. The TLV is a recommendation by ACGIH, with only a guideline status. As such, it should not be confused with exposure limits having a regulatory status, like those published and enforced by the Occupational Safety and Health Administration (OSHA). Many of the OSHA regulatory exposure limits are based on the 1968 ACGIH TLVs, adopted into Federal regulation in that year under the Walsh-Healy Public Contract Act, and so are over 40 years old. Many are not considered by the industrial hygiene community to be sufficiently protective levels. The National Institute of Occupational Safety and Health (NIOSH) publishes RELs (recommended exposure limits) which OSHA takes into consideration when promulgating new regulatory exposure limits.

Definitions

The TLV for chemical substances is defined as a concentration in air, typically for inhalation or skin exposure. Its units are in parts per million parts of air (ppm) for gases and in milligrams per cubic meter (mg/m^3) for particulates such as dust, smoke and mist.

The basic formula for converting between ppm and mg/m^3 for gases is ppm = (mg/m^3) * 24.45 / molecular weight. This formula is not applicable to air-borne particles.

Three types of TLVs for chemical substances are defined:

1. Threshold Limit Value-Time Weighted Average (TLV-TWA): average exposure on the basis of a 8h/day, 40h/week work schedule
2. Threshold Limit Value-Short Term Exposure Limit (TLV-STEL): Spot exposure for a duration of 15 minutes, that cannot be repeated more than 4 times per day.
3. Threshold Limit Value-Ceiling (TLV-C): Absolute exposure limit that should not be exceeded at any time.

There are TLVs for physical agents as well as chemical substances. TLVs for physical agents include those for noise exposure, vibration, ionizing and non-ionizing radiation exposure and heat and cold stress.

Similar Concepts

The TLV is equivalent in spirit (although the materials covered, values recommended, and definitions used can differ) to the following concepts.

- WEEL (*Workplace Environmental Exposure Levels) created by a committee of the American Industrial Hygiene Association (AIHA).*

Similar concepts used in other countries include:

- Australia
  - OES (Occupational Exposure Standard)
- France
  - VME (*Valeur Moyenne d'Exposition*)
  - VLE (*Valeur Limite d'Exposition*)
- Germany
  - AGW (*Arbeitsplatzgrenzwert*)
  - MAK (*Maximale Arbeitsplatz-Konzentration*)
- Netherlands
  - MAC (*Maximaal Aanvaarde Concentratie*)
- Malaysia
  - PEL (*Permissible Exposure Limits*)
- Poland
  - NDN (*Najwyzsze Dopuszczalne Natezenie*)

Chapter 8

Environmental Monitoring

(a) Chemical Monitoring

Toluene (Toxicology)

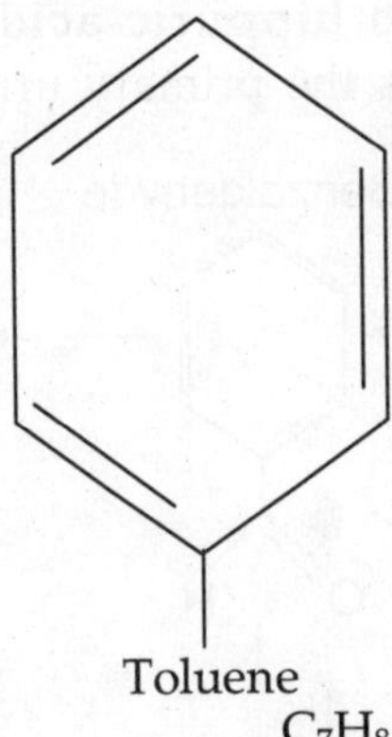

Toluene

Chemical formula	C_7H_8
Molecular weight	92.14052
Excretion	Lungs
Routes of Exposure	Skin, GIT, Lungs

Metabolism in Humans

While a significant amount of toluene, 25%-40%, is exhaled unchanged via the lungs, a greater proportion is metabolised and excreted via other pathways. The primary route of toluene metabolism is by hydroxylation to benzyl alcohol by members of the cytochrome P450 (CYP) superfamily. There are five CYPs which are important in toluene metabolism, CYP1A2, CYP2B6, CYP2E1, CYP2C8, and CYP1A1. The first four seem to be involved in the hydroxylation of toluene to benzyl alcohol. CYP2El seems to be the primary enzyme in the hydroxylation of toluene, accounting for roughly 44% of toluene metabolism; however, there is a great deal of ethnic variability, in the Finnish population, for example, the primary enzyme is CYP2B6. CYP2E1 catalyses the formation of benzyl alcohol and p-cresol, while CYP2B6 produces comparatively little p-cresol.

It is believed that in humans, benzyl alcohol is metabolised to benzaldehyde by CYP rather than alcohol dehydrogenase; however, this belief does not appear to be universal. Benzaldehyde is in turn metabolised to benzoic acid, primarily by mitochondrial aldehyde dehydrogenase-2 (ALDH-2), while only a small percentage is metabolised by cytosolic ALDH-1.

Benzoic acid is metabolised to either benzoyl glucuronide or hippuric acid. Benzoyl glucuronide is produced by the reaction of benzoic acid with glucuronic acid, which accounts for 10-20% of benzoic acid elimination. Hippuric acid is also known as benzoylglycine and is produced from benzoic acid in two steps: first benzoic acid is converted to benzoyl-CoA by the enzyme benzoyl-CoA synthase; then benzoyl-CoA is converted to hippuric acid by benzoyl-CoA:glycine N-acyltransferase. Hippuric acid is the primary urinary metabolite of toluene.

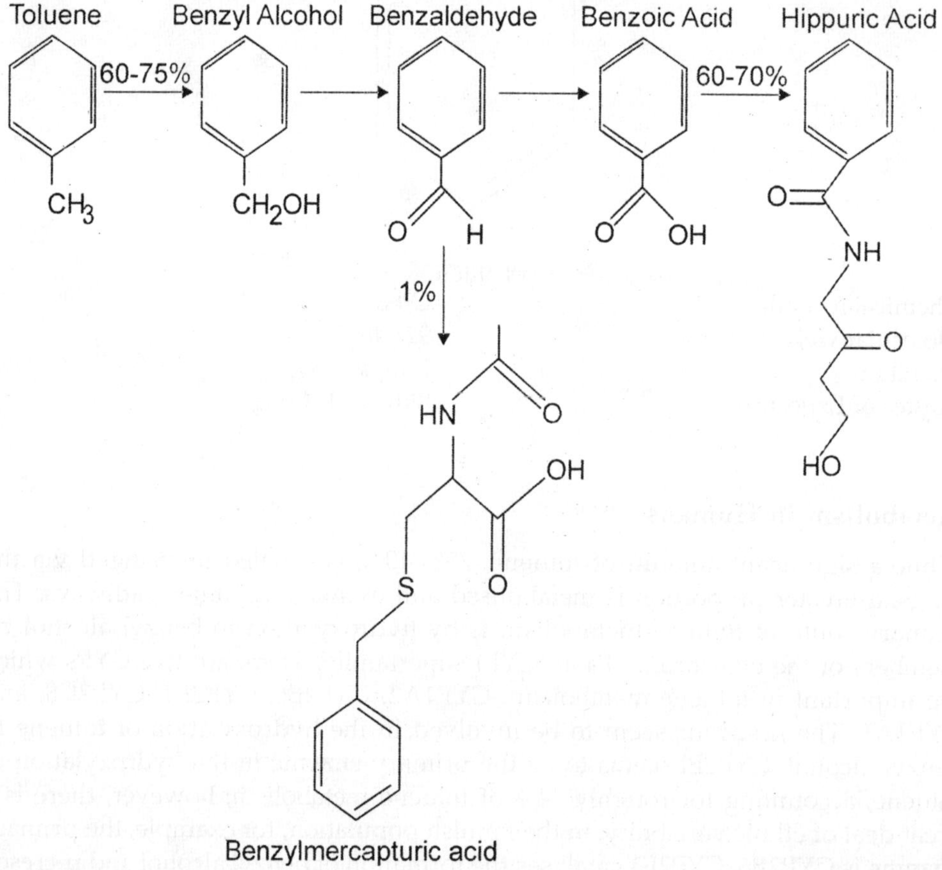

Toluene Benzyl Alcohol Benzaldehyde Benzoic Acid Hippuric Acid

60-75% 60-70%

CH_3 CH_2OH O H O OH O NH O HO

1%

HN O OH S O

Benzylmercapturic acid

Ring hydroxylation to cresols is a minor pathway in the metabolism of toluene. The majority of the cresol is excreted unchanged in urine; however, some of the p-cresol and o-cresol is excreted as a conjugate. Studies in rats have shown that p-cresol is primarily conjugated with glucuronide to produce p-cresylglucuronide, though this may not be applicable to humans. o-Cresol appears to be excreted mostly unchanged in urine or as the glucuronide or sulfate conjugate. There appears to be some dispute over whether m-cresol is produced as a metabolite of toluene or not.

OH
1%
CH
CH
1%
m-cresol
0.1%
OH
OH
CH
CH
o-cresol
p-cresol

Environmental Influences

When exposure to toluene occurs, there is usually simultaneous exposure to several other chemicals. Often toluene exposure occurs in conjunction with benzene and since they are to some degree metabolized by the same enzymes, the relative concentrations will determine their rate of elimination. Of course, the longer it takes for toluene to be eliminated, the more harm it is likely to do.

The smoking and drinking habits of those exposed to toluene will partially determine the elimination of toluene. Studies have shown that even a modest amount of acute ethanol consumption can significantly decrease the distribution or elimination of toluene from the blood resulting in increased tissue exposure. Other studies have shown that chronic ethanol consumption can enhance toluene metabolism via the induction of CYP2E1. Smoking has been shown to enhance the elimination rate of toluene from the body, this perhaps occurs as a result of enzyme induction.

The diet can also influence toluene elimination. Both a low-carbohydrate diet and fasting have been shown to induce CYP2E1 and as a result, increased toluene metabolism. A low-protein diet may decrease total CYP content and thereby reduce the elimination rate of the drug.

Measure of Exposure

Hippuric acid has long been used as an indicator of toluene exposure; however, there appears to be some doubt about its validity. There is significant endogenous hippuric acid production by humans; which shows inter- and intra-individual variation influenced by factors such as diet, medical treatment, alcohol consumption, etc. This suggests that hippuric acid may be an unreliable indicator of toluene exposure. It has been suggested that urinary hippuric acid, the traditional marker of toluene exposure is simply not sensitive enough to separate the exposed from the non-exposed. This has led to the investigation of other metabolites as markers for toluene exposure.

Urinary *o*-cresol may be more reliable for the biomonitoring of toluene exposure because, unlike hippuric acid, *o*-cresol is not found at detectable levels in unexposed subjects. o-Cresol may be a less sensitive marker of toluene exposure than hippuric acid. o-Cresol excretion may be an unreliable method for measuring toluene exposure because o-cresol makes up <1% of total toluene elimination.

Benzylmercapturic acid, a minor metabolite of toluene, is produced from benzaldehyde. In more recent years, studies have suggested the use of urinary benzylmercapturic acid as the best marker for toluene exposure, because: it is not detected in non-exposed subjects; it is more sensitive than hippuric acid at low concentrations; it is not affected by eating or drinking; it can detect toluene exposure down to approximately 15 ppm; and it shows a better quantitative relationship with toluene than hippuric acid or *o*-cresol.

Effects of Long Term Exposure

Serious adverse behavioural effects are often associated with toluene abuse related to the deliberate inhalation of solvents. Long-term toluene exposure is often associated with effects such as: psychoorganic syndrome; visual evoked potential (VEP) abnormality; toxic polyneuropathy, cerebellar, cognitive, and pyramidal dysfunctions; optic atrophy; and brain lesions.

Toluene causes postural tremors by increasing extracellular concentrations of γ-aminobutyric acid (GABA) within the cerebellar cortex. Treatment with GABA agonists such as, benzodiazepines provide some relief from toluene induced tremor and ataxia. An alternative to drug treatment is vim thalamotomy. The tremors associated with toluene misuse do not seem to be a transient symptom, but an irreversible and progressive symptom which continues after solvent abuse has been discontinued.

There is some evidence that low-level toluene exposure may cause disruption in the differentiation of astrocyte precursor cells. This does not appear to be a major hazard to adults; however, exposure of pregnant women to toluene during critical stages of fetal development could cause serious disruption to neuronal development.

(b) Biological Monitoring

BIOLOGICAL MONITORING OF PESTICIDE

Pesticide Residue

The used pesticides and their degradation products and metabolites in various compartments (air, water, soil and biota etc) are known as pesticide residue which ultimately may find way to human body through food chains causing various health hazards which is known and well documented.

Tolerance Limits of Pesticide in Food/Water

Maximum residue level (MRL) or tolerance limit mg/kg (ppm) of pesticide are legal limits for concentration of specific compounds or specific food which is being consumed by human being for longer period is not harmful.

Pesticide Residue Analysis

Inspite of the fact that pesticide are effective arms in the hands of the farmers against pests, their continuous use may leave residue in soil, water and food which have toxic effect on plants, animals and human beings. Therefore, pesticide residue analysis is an important step in environmental and biological samples to ensure safety and protection from health hazards.

Cleaning of Glasswares

The most important to be considered in pesticide residue analysis is the cleaning of glassware because if they are used in different steps of analysis then contamination can come from that. The cleaning of glassware may be done by the following steps to make them entirely free from contamination.

(a) Soaking and washing in a high temperature 50^{0}C bath of synthetic detergents in water.

(b) Rinsing with tap water.

(c) Rising with distilled water.

(d) Finally before being used with those solvent going to be used for extraction.

Cleaning glassware used to concentrate sampling KD flask or conical flask should include a soak for atleast 15 minutes in hot (40-50^{0}C) chromic acid cleaning solution after tap water rinsing to remove all traces of organic material. This soak is followed by thorough rinsing with tap and distilled water, then with acetone and hexane.

Sample Collection

Representative Vs. Biased Sampling

It is important that the analyst be aware of representative sample and he being consulted when the sample is to be collected is decided so that it is valid for the purpose of analysis and valuable time is not wasted on a worthless sample.

Sample Container

The container should be plastic or metallic, it should be pre-washed and made up of glass.

(a) Wide mouth glass bottle of teflon or aluminium foil lined screw cap for tissues (autopsy) of less than 25g.

(b) Glass vials of at least 7 ml capacity (avoid rubber or cork caps) for blood.

(c) Empty pesticide-grade solvent bottles for water effluents.

(d) Glass jars for environmental or agricultural samples.

Specimen included for organochlorine compound analysis are never wrapped in paper, cardboard or plastic.

Sample Composition

After collection of valid gross sample, compositing or reduction to analytical size sample may be required, especially for agricultural and food samples. The general requirement is that the small analytical sample must be fully representative of the gross sample collected.

Storage of Samples

As a general rule, sample should be analyzed as soon as possible, after collection. If the storage is necessary , it should be under prescribed conditions that preserves the integrity of the original sample.

Fish tissue	to be extracted within 24 hours	Refrigerated Temperature (2 to 4°C)
	If not extracted within 24 hours	deep frozen (-12 to -18°C)
Blood	If analysed within 3 days	Refrigeration temperature
	(separate serum immediately)	-2 to 4°C
	if not analysed within 3 days	deep freeze (-12 to -18°C)
Agricultural	Process extraction immediately	store in deep freeze -17°C

Environmental if not possible to extract immediately store in dark at 4°C

If lengthy storage is required prior analysis, a good alternative to storage sample is to extract the sample at once, remove most or all of the solvent and store extract at a low temperature. Decomposition in samples stored can be evaluated by storing spiked control along with the sample.

Extraction

Many solvents are employed for extracting residues depending on the polarity of the pesticide and the amout of co-extractive expected from the particular substrate.

Solvent to be used with increasing polarity

Hexane (petroleum ether)	least polar
Benzene	
Ethyl ether	
Methylene chloride	
Ethyl acetate	
Acetonitrile	
Methanol	more polar

CC– – – – – – – – – – –→OP– – – – – – – – – – – – –→Carbamate

Non-polar/semi-polar Polar

Hexane/petroleum ether– – – – – – – – – – – –→dichloromethane

Chloroform, diethyl ether, ethyl acetate, benzene, acetonitrile, methanol, acetone in 2 or 3 component mixtures have been widely used

- Acetonitrile is an excellent general purpose extraction solvent for low fat content samples
- Hexane/acetonitrile systems are widely recommended for partition clean up.

The extraction of pesticides from biological tissue will be governed by:

- nature of analyte
- toughness
- water content
- lipid content

Different techniques are employed for bringing the extraction solvent and sample into contact

Pesticide present in matrix

Surface extraction by simple Internal residue washing procedure with solvent

Macerate		Fine particle sample
Blending (e.g. Plant	Soxvlet	Partioning
Material, biological	(soil/sediment)	(water/waste water)

Water is the sample may give rise to emulsions with non polar solvents avoided by using a drying agent such as Na_2SO_4 or propanol (individual or together).

Meat or muscle samples have too much connective tissue which can break down by Using formic acid.

Grinding the sample with Na_2SO_4 and sand followed by heating on water bath with solvent.

Note: Disadvantage –More volatile pesticide eg lindane might be lost in the way

Clean Up

Injection of extract into a GLC, HPLC can cause extraneous peaks, deterioration of peak resolution and column efficiency and loss of detector sensitivity and can greatly shorten the lifetime of expensive columns. It requires clean up.

Partitioning between immiscible solvents

Adsorption chromatography Alumina are

1. Florisil
2. Silica gel

Gel permeation chromatography (to separate OC and PCB)

Chemical destruction of interfering substances with acid, alkali or oxidising agents

Distillation

Adsorption chromatography

Most materials can be activated by strong heating and some may be activated for a particular purpose by penetrating with an acid or base (alumina) or an organic solvent (charcoal) deactivation of a polar adsorbent to a desired level has been achieved by addition of certain percentage of water.

Florisil has proven to be nonuniform in elution characteristics and therefore each batch requires careful presenting of adsorptive properties prior to use.

Elution pattern may depend upon

- Quality of florisil
- Packaging procedure
- Temperature
- Humidity
- Amount of adsorbent
- Polarity of solvent

One should use the least polar solvent that will elute the pesticide from adsorbent to minimize the coelution of polar impurities.

Concentration

Purified extracts of elute solvents containing even somewhat volatile compounds are concentrated with minimum losses to a volume of 5 to 10 ml using following devices:

(a) Kuderma-Danish evaporator

(b) Rotary evaporator

(c) Nitrogen evaporator apparatus.

It is important to avoid pesticide loss or decomposition during evaporation steps using a keeper, such as a paraffin helps to retain the compounds and greatly reduces loss. However, the keeper may interfere with some analysis, especially by FID or spectrometry.

Evaporation to dryness should never occur. If the complete removal of particular solvent is required, solvent exchange can be carried out so that the samples never gets to dryness.

Satisfactory recoveries are obtainable when the residue levels are relatively high, but significant losses have been documented of even the more stable pesticide at low concentration levels.

End Analysis

In a well cleaned extract, there are a number of other impurities which may interfere during analysis. Hence identification of trace residues is a major requirement of pesticide analyst. The identification of pesticide residues should always be confirmed by a method different from that used in the initial determination since interpretation of results as well as reliable quantification depend on correct identification. Multiresistant GC analytical method do not provide irrefitable identification since interfering materials and atifacts are often observed and metabolic and decomposition products are encountered.

The difficulty in identification arises in sample where very low quantity of pesticides are present. Generally useful confirmatory method (e.g. IR) require great quantity and purity. Following points are to be considered for choosing the method for confirmation:

- Nature of pesticide
- Presence of other residue
- The level of pesticide
- The type of pesticide
- The type and amount of sample.

The following analytical techniques are being used to identify and quantify the pesticide content in residue analysis.

Confirmation by GC retention time from polar and non-polar column, TLC RF value, paper chromatography RF value, GC retention time of a derivative formed by chemical/photochemical reaction, p-values in different solvent pairs, high performance liquid chromatography, spectrophotometric method IR, NMR, MAS and MASS coupled with GC/LC.

Preparation of Standards

Use the standard chemical as a reference material of the purity more than 99% and follow the following recommendations for storage of pesticide analytical reference standard.

* Select a solvent, such as isooctane or toluene, that will dissolve the standard material and evaporate slowly.
* Store the standard in relatively large volume solution (50-100 ml) to reduce the percentage volume losses to acceptable levels.
* Monitor the solvent loss by placing a mark on the side of the solution container and then discard the solution when 3-5% solvent has evaporated.
* Select a container, such as screw cap precipitation bottle or a large volume volumetric flask that will not allow rapid solvent evaporation.
* Store the standards away from light in refrigerator when not in use to reduce evaporation and reference material decomposition.
* Replace the easily decomposed reference standard solutions at the recommended intervals.
* Check periodically solutions by comparison against fresh dilutions of the stock solution for validity of reference standard.
* Do not store any standard solutions for longer than one year.

Gas Liquid Chromatography

The equipment is mainly used for quantification and identification of pesticide residues. The equipment has the following parts:

Column

1. GC column tubes are made of borosilicate glass. Copper and stainless steel should not be used because they cause decomposition of compounds.
2. Column efficiency and peak resolution depend upon the number of theoretical plates. A 6′column of 3000 plates will usually provide acceptable resolution mixtures encountered to residue analysis.
3. Lowloading of liquid phase beneficially effects efficiency. The minimum coating that can be used is limited to the amount for complete coverage of the support, usually 1-3%.

Commonly used liquid phases in residue analysis

Liquid phase	Chemical name	Equivalent	Max temp used
DC-200	Methyl silicone	OV-1, OV-10, SE-30, SP-210	350^{0}C
QF-1	Trifluoropropylmethyl	OV-210, SP-2401	275^{0}C
SE-52	Methyl silicone (10% phenyl substituted)	OV-3	350^{0}C
XE-60	Cynoethyl methyl silicone	Ov-225	250^{0}C
OV-17	Methyl silicone (50% phenyl substituted)	SP-2250	375^{0}C
OV-7	Methyl silicone (20% phenyl substituted)	-	-
DEGS	Diethylene glycol	Succinate	–

Column Stability

When a succession of dirty extracts are passed through the system, the column performance is usually affected. The most prevalent symptoms of injection over loading are:

a) depressed peak

b) lowered efficiency and resolution

c) on-column breakdown of precipitates

d) erratic recoveries

e) unsymmetrical peaks.

Resistance to on-column compound decomposition

Unless a column is properly prepared, conditioned and maintained, it can cause such compounds on ending and/or pp-DDT to undergo some degree of decomposition.

Column performance with age

The older column is promoting degradation of DDT to DDD and retention time have lengthened. Daily monitoring of DDT breakdown is important for early indication of contamination of injection port and/or column.

1. Detector
2. Characteristic of detector
3. Sensitivity
4. Linear dynamic range
5. Selectivity

Types

a) Electron capture detector (ECD): specific and highly sensitive for halogenated compounds.

b) Flame ionization detector (FID) : general detector for organic compound except formic acid.

c) Thermal conductive detector (TCD): for all organic compounds and permanent gases.

d) Flame photometric detector (FPD) : Highly sensitive for sulfur.

e) Alkali flame ionization detector (AFID): Very enhanced response to phosphorus.

f) Nitrogen phosphorus detector (NPD) : Involving an electricity heated bead which require reduced flow of hydrogen and temperature. An increased response to 'N' and high sensitivity to 'P'.

Analysis of Heavy Metals in Water and Waste Water

The metals are categorized in two depending on their nature, i.e. non-volatile metals (Cd, Cu, Cr, Fe, Mn, Ni, Pb, Zn etc.) or volatile metals (Hg, As etc.). Volatile metals can be analyzed on flame AAS or ICP while non-volatile metals are analyzed on AAS or ICP equipped with a Vapor Generation Assembly (VGA). The processing of the samples for these categories are different.

Principle of the Analytical Instruments

Atomic Absorbtion Spectrophotometer (AAS)- This is one of the most volatile and sought after method for quantification of metals. In AAS a light beam is directed through the flame, into a monochromator, and on to the detector that measures the amount of light absorbed by atomised element in the flame. The amount of energy at the characteristic wave length absorbed in the flame is proportional to the concentration of the element in the sample over a limited concentration range.

Inductively Coupled Plasma Atomic Emission Spectrophotometer (ICP)- An ICP source consists of flowing stream of argon gas ionized by an applied radio frequency field typically oscillating at 27.1 MHz. The field is inductively coupled in the ionized gas by a water cooled coil sorrounding a quartz torch that supports and confines the plasma. A sample aerosol is generated in an appropriate nebulizer and spray chamber and is carried in to plasma through an injector tube located within the torch. The sample aerosol is injected directly into the ICP subjecting the constituent atoms to temperatures of about 8000-10000^{0}K, which results in complete dissociation of molecules. Ionization of high percentage of atoms produces ionic emission spectra. The efficient excitation provided by ICP results in low detection for many elements over a wide concentration range.

Equipments and Chemicals Needed

Equipments

1. Atomic Absorption Spectrophotometer with appropriate Hollow Cathode lamps
2. Inductively Coupled Plasma Atomic Emission Spectro photometer

Chemicals and reagents:

1. Nitric acid (HNO_3) Concentrated AnalR grade
2. Perchloric acid ($HclO_4$) 70% AnalR Grade
3. Metal standards (Stock solutions)

They are available in 1000 ppm concentrations and can either be procured from reputed firms like Merck, Sigma or any other make or can be prepared in lab using high purity (99.9%) metal powders or their respective salts.

(i) Cadmium (Cd) : Dissolve 0.1gm Cd metal in 4ml concentrated HNO_3 and dilute to 1000 ml with deionized water.

(ii) Chromium (Cr) : Dissolve 0.1 g of Cr metal or 0.1923 of CrO_3 in water. When dissolution is complete acidify with 10 ml of concentrated HNO_3 and dilute to 1000 ml with deionized water.

(iii) Copper (Cu) : Dissolve 0.1 g of copper metal in 2.0 ml of concentrated HNO_3 and 10 ml of concentrated HNO_3 and dilute to 1000 ml with deionized water.

(iv) Iron (Fe) : Dissolve 0.1 gm of iron metal in mixture of 10 ml of 50% HCl and 3ml of concentrated HNO_3 and dilute to 1000 ml with deionized water.

(v) Manganese (Mn) : Dissolve 0.1 g manganese metal in 10 ml concentrated HCl mixed with 1 ml of concentrated HNO_3 and dilute to 1000 ml with deionized water.

(vi) Lead (Pb) : Dissolve 0.1g of lead metal in 5 ml of aquaregia or 0.1598g lead nitrate in minimum amount of 50% HNO_3, add 10 ml HNO_3 and dilute to 1000 ml with deionized water.

(vii) Zinc (Zn) : Dissolve 0.1 g of Zinc metal in 20 ml of 50% HCl and dilute to 1000 ml of deionized water.

Processing of the sample for non-volatile metals:

1. Mix the sample thoroughly and transfer a suitable volume (200 ml) into suitable flask or beaker.
2. Add 5ml of concentrated HNO_3 and a few boiling chips and evaporate on a hot plate to 10-15 ml.
3. Add 5 ml of each concentrated HNO_3 and $HClO_4$ after cooling the flask.
4. Evaporate the hot plate till dense white fumes of perchloric acid appear.

5. If the sample has turned brown, more concentrated HNO_3 can be added and the process of heating repeated for complete digestion.
6. Cool and dilute to about 20 ml with deionized water and boil to expel chlorine, cool and filter.
7. Transfer the contents quantitatively to a 25 ml volumetric flask and make up.
8. The above digested sample is aspirated for quantification on an AAS or ICP.
9. Appropriate blanks using the same volume of deionized water are also processed in the same manner.

Precautions to be Taken in Sampling and storage

Always use high polythene containers. These containers should be cleaned with 10% HNO_3 and then with distilled water. Preserve the sample immediately after sampling by acidifying with concentrated HNO_3 to pH< 2. Usually 1.5 ml concentrated HNO_3/l is sufficient for short-term preservation. The samples should be stored at 4⁰C till the time of processing and analysis.

Processing of Sample for Mercury Analysis

The mercury in samples is reduced to elemental state by the processing and then volatized in the presence of sodium of sodium borohydride in an acidic medium. The later reaction takes place in VGA of the AAS. These volatized mercury vapors pass through the absorption cell of AAS in the light path (radiation at 253.7 nm by Hg hollow cathode lamp). The sensitivily of the instrument is < 1ppb (1ì/l). The absorption cell for mercury has quartz windows on both sides and the flame is OFF.

Alternatively mercury can be analyzed on cold vapor-mercury analyzer using $SnCl_2$ (20% in distilled water) as reductant and 10% HNO_3 for acidification of sample. The sensitivity of this instrument is 20 ng (in total volume of sample).

Standard mercury (Hg) stock solution : Dissolve 0.1354 g $HgCl_2$ in about 70 ml of distilled water and 1ml of concentrated HNO_3 is added and volume made up to 100 ml with distilled water, (this gives the stock solution of 1000 ppm). Working solutions are prepared fro this stock by dilution to contain 1-20 ng Hg/ml or 1-20ìg/l.

Reagents

1. Concentrated hydrochloric, nitric and sulfuric acid Analytical Grade.
2. Potassium permanganate solution: Dissolve 50g $KmnO_4$ in water and dilute to 1000 ml distilled water.
3. Hydroxylamine hydrochloride solution: Dissolve 10 g Hydroxylamine hydrochloride ($NH_2OH.HCl$) in 100 ml distilled water.

4. Sodium borohydride solution: Dissolve 0.3 g sodium borohydride in 100 ml of 0.5% NaOH solution. This solution should be prepared fresh everyday.

Procedure

1. Transfer 150 ml of water sample in a 250 ml conical flask.
2. Add 5 ml of HNO_3 and 10 ml of 50% sulfuric acid to the sample.
3. Mix thoroughly and add 15 ml of $KMnO_4$ solution. Let stand overnight.
4. Add 1ml of hydroxylamine hydrochloride solution so that the colour of the $KMnO_4$ disappears.
5. Make up the volume of the sample to 100 ml with distilled water.
6. This sample is used for the analysis of mercury on AAS-VGA.

Analytical method:

1. Warm the hollow cathode lamp in the AAS.
2. Set up the absorption cell I in the light path.
3. Set the instrument to 5.0 sec read time and 40 sec delay time.
4. Put 6N HCl in the container meant for acid channel of VGA.
5. Put sodium borohydride (0.3% in 0.5 NaOH) for the reductant channel of VGA.
6. Aspirate 1% HCl blank in distilled water to set the instrument ZERO.
7. Plot a curve using 1,5,10 and 20ppb (ng/ml) of mercury standards.
8. Thoroughly wash the sample tube with distilled water so that the instrument again reads ZERO and then aspirate the sample.
9. Take the mean Hg reading of the sample (in ppb).

Precautions to be taken in the sampling and storage of mercury:

Samples would be preferably collected in amber colored glass bottles precleaned with 10% nitric acid. As mercury can readily be lost from the samples, it has to be preserved by the addition of 2 ml/l of 20% $K_2Cr_2O_7$ in 2% HNO_3 to pH<2.

Processing of sample for Arsenic analysis:

Arsenic is analyzed on AAS equipped with VGA with a special absorption cell with both the sides open and the instrument is set in flame mode. Arsenic hydrides generated in the system dissociate in the heated absorption cell thus freeing the arsenic atoms which cause the absorbance of the energy from the arsenic hollow cathode lamp.

Alternatively, as can also be estimated in a Gutzeit's apparatus where the sample is reacted with nascent hydrogen generated by a reaction of zinc powder with HCl to produce arsenic gas in the presence of KI. This gas is reacted with silver diethydithiocarbamate to form red complex, which can be determined calorimetrically at 535 nm. The sensitivity of this method is 1μg in the whole sample in the reaction flask.

Standard Arsenic stock solution:

Dissolve 1.320 g of arsenic trioxide in 10 ml of distilled water having 4N NaOH and make the volume to 1 litre. This will give 1000mg As/1 (stock solution). This stock solution can be diluted to give 1,5, 10 and 20ng As/ml. For use on AAS-VGA the standards should be made in 6N HCl (final concentration) and 10 ml of 10% KI solution should be added per 100 ml of standard 30 mn prior to analysis.

Reagents

1. Hydrochloric acid AnalR Grade 10 N.
2. Sodium borohydride solution : Dissolve 0.6 g sodium borohydride in 100 ml of 0.5% NaOH solution. This solution must be prepared fresh every day.
3. 10% Potassium iodide (KI) in distilled water.

Procedure

1. Transfer 50 ml of water sample in a 250 ml conical flask.
2. Add 40 ml of concentrated HCl and 10 ml of KI solution to the sample.
3. Mix thoroughly and allow to stand for 60 min at room temperature.
4. This sample is used for the analysis of arsenic on AAS-VGA.

Analytical Method:

1. Warm the Arsenic Hollow cathode lamp in the AAS.
2. Set up the absorption cell (with open sides) in the light path.
3. Set the instrument to 5.0 sec read time and 40 sec delay time. Set it to take triplicate readings.
4. Put 10N HCl in the container meant for acid channel of VGA.
5. Put sodium borohydride (0.6% in 0.5N NaOH) for the reductant channel of VGA.
6. Aspirate appropriate blank in distilled water to set the instrument ZERO.
7. Plot a curve using 1, 5, 10 and 20 ppb (ng/ml) of arsenic standards.
8. Thoroughly wash the sample tube with distilled water so that the instrument again reads ZERO and then aspirate the sample.
9. Take the mean as reading of the sample (in ppb).

Calculation of the results:

Metal concentration in mg/ml (in ppm) or ng/ml (in ppb for Hg and As) =

RS-RB X make up volume of the sample (ml/initial volume of sample (ml)

Where:

RS- Reading of sample (in ppm or ppb)

RB- Reading of the processed blank (in ppm or ppb)

ENUMERATION OF BLUE-GREEN ALGAE (CYNOBACTERIA) IN WATER

Eutrophication of water bodies caused due to diverse types of chemical and organic contaminants from extraneous sources tend to form cynobacterial blooms/scums on the surface of open water systems and other microorganisms in different layers of water columns. The epilimnetic primary production rate of cynobacterial bloom cause certain nuisance as depleting hypolimnetic oxygen level needed for survival of aquatic fauna and cease the growth of other micro plant species. Not all these, after lysis of living or dead cells in cynobacteria in late summer and early autumn season, produce secondary metabolites (organic compounds) which have unique structure and both poisonous and non-poisonous properties. The poisonous secondary metabolites are widely known as cynobacteria toxins comprised into hepatotoxins, cytotoxins, neurotoxins and endotoxins which cause deleterious effects like hepatotoxic, cytotoxic, neurotoxic and endotoxic in animals and human beings finally resulting into death.

Toxic cynobacteria (blue-green algae) found in eutrophic, municipal and residential water supplies are in increasing environmental hazard in several parts of the world. Cynobacteria produce lethal toxins and domestic and wild animal deaths have been caused by drinking water contaminated by these toxins. Among the species causing death in livestock, *Microcystis aeruginosa* blooms are the most common. Toxins producd by cynobacteria include alkaloid neurotoxins and peptide hepatotoxins, viz., microcystins and nodularin and are water-soluble and temperature stable. Since these organisms occur in both recreational and drinking water lakes and rivers and are known to be strong liver tumor promoter, they represent a health hazard to human populations.

Enumeration

Enumerating blue-green algae is a time-consuming exercise depending on the level of accuracy and the level of toxic identification required. The basic procedure for counting blue-green algae begins with the correct handling of the sample on the arrival at the laboratory. If the same day live counting is to be performed on the sample, the sample needs to be refrigerated until concentration or examination is to begin. If it is not to be examined the sample on the same day it has been collected, the sample needs to be preserved with lugol iodine. To preserve a sample with lugol, you need to add 0.7 ml of lugol to 100 ml of sample and refrigerate the sample until concentration and/or examination. If the algae load has a cell density of less than 500 units per ml the sample will require concentrating.

Concentration Methodologies

Algal concentration methods are presented in the standard methods for the examination of water and waste water (APHA, 2000).

Lugol Sedimentation

This method is based on the algae present in the same sample being sedimented to the bottom of a vessel (A measuring cylinder is commonly used) for a period of time. After sedimentation, the top 90% of volume is carefully siphoned-off without distributing the sedimented algae. A subsample of the remainder of the sample is then used for enumeration.

Membrane filtration

This method is based on a volume of sample being filtered through a membrane of pore size 0.45 µm. The algae are trapped into the membrane and then gently brushed off and resuspended in distilled water. A subsample of this is again used for enumeration.

Centrifugation

This method is based on centrifuging the sample and decanting of the top volume. The algae is then resuspended and a subsample is used for enumeration. Other non-standard methods are also used for concentration of algae samples and include sand filtration. Sand filtration is based on filtering the sample through a plug of fine sand and then resuspending the algae trapped on the sand into distilled water. A subsample of this is then used for enumeration. The method of choice for concentration of algal samples depends on the needs and the program objectives.

Method Choice

Results obtained from validation studies conducted at laboratory indicates that each method of concentration has its advantages and disadvantages. These should be taken into consideration when choosing a method.

Performance of Enumeration

The prepared sample is allowed to come to room temperature before dispensing a 1 ml subsample into a counting cell. A glass Sedgwick Rafter changer is commonly used. The sample should not be disturbed for atleast 15 minutes before examination so that algae may settle to the bottom of the changer. The sample is then placed under the microscope and a number of fields are examined. The number of random fields or strips to be examined depends upon the algal density and precision required. It is recommended that 30 fields be counted within the chamber so as to include 90-95% of species present, and 100 units of the most common algae taxa counted. Laboratory also stipulates that 100 fields be counted for very low algae density samples. The cells within each trichome are to be counted where possible. Where it is impossible to determine individual cells within a trichome, due to their small size and/or indistinguishable cell divisions, count the trichome as one unit. The cells within each colony are to be counted (or estimated). In case of counting cells which are dividing, consider those in which the seperating wall can be clearly discerned to be two cells. A convention needs to be followed for cells or units lying on the boundary lines in the S-R chamber. When a trichome of colony extends outside the field slightly (less than one-half but more than one-quarter) do not count the cells outside the field. When a trichome or colony extends into the field being examined by less than a quarter, do not count this trichome or colony.

Recording of Data

It is an ideal to have standard algae recording sheets containing information such as laboratory identification number, sample reference number and sample description, counters name and data counted, volume of sample used for concentration, area counted and factors to calculate final cell count. The counter should record the number of colonies, and if possible, the number of cells in each unit for each blue green algae genus or species observed. Specifically designed counting software is now available which cuts down on data entry time and some will calculate results. The number of units and cells per ml for each genus or species counted is calculated. Then calculate the number of cells per unit and then adjust for any concentration or dilution of the sample. These results are then commonly transferred to an electronic data base for storage and reporting. Quality assurance programmes, which encompass documentation, are an essential part of any laboratory procedures, and serve to create component operators and reliable results in algae enumeration.

VARIOUS METHODS FOR ENUMERATION OF BACTERIA IN WATER

Bacteria are present everywhere in most of the biotic and abiotic components of atmosphere including underground strata. Natural commodities such as air, water, soil and raw food items, essential for life, are usually contaminated with a variety of bacterial species and some of which are pathogenic to human beings and other animals. The pathogenic bacteria causes several water-borne diseases among which gastroenteritis, cholera, dysentery and typhoid are most common. The causative bacteria of these diseases are *Escherichia coli, Vibrio cholerae, Shigella dysenteriae and Salmonella typhimurium* respectively. The degree of bacterial contamination in air, water, soil and food including any liquid, solid, swab, biological and clinical samples can be determined by enumeration of test bacteria by several methods.

Since bacteria are microscopic organisms these can be enumerated by microscopic and culture methods. Further, as the microscopic counting of bacteria is too tedious, only culture techniques are used for counting the bacteria. For the assessment of bacteriological quality of water enumeration of certain pollution indicator bacteria or group of bacteria, viz., coliform, faecal coliform and faecal strefitococci, is internationally recommended. Culture techniques requires selective solid or liquid medium, incubation period and temperature along with standard laboratory conditions, instruments and facilities necessary for the microbiological work.

CLEANING GLASSWARE

A good quality detergent powder or liquid is used for cleaning glasswares. Chromic acid or any other organic solvent, e.g. alcohol, etc., may be applied carefully for difficult cleaning problems. Glasswares are sufficiently washed with warm water and then running water to remove detergent and acid. After cleaning, the glasswares are kept inverted in slotted plastic, non-rusting metallic or wire gauze basket or tray to drain down the water and dried properly. Glassware with creck or any break are discarded.

STERILIZATION

1. Glasswares

All the clean and dried sampling bottles (glass stopered), pipettes and petridishes are sterilized in a hot air oven at 170 + 10^{0}C for one hour. Sampling bottles for collecting the chlorinated water sample contain 0.1 ml of 8% (w/v) solution or a crystal of sodium thiosulfate for each 100 ml of bottle capacity before sterilization to neutralize the residual chlorine in water sample. The mouth with neck of all the sampling bottles is wrapped with kraft paper and tied with cotton twine before sterilization. The mouth piece of all 10-1ml pipettes are cotton wool plugged and kept in a non-rusting metallic drum with cover, putting tip of pipettes towards bottom. Petridishes are kept in pairs and piled in groups of 4-6 pairs, in clean rustless metallic box.

2. Culture Media, Dilutents and Solutions

Most of the broth as well as agar culture media, diluent and solutions are sterilized in autoclave or pressure cooker (upto 3 whistles) under steam pressure of 15lbs/inch^2 (121^{0}C) for 15 minutes. Certain high temperature sensitive media, diluents or solutions are sterilized under steam without pressure or by boiling for 30 minutes. Some heat sensitive media or solutions can be sterilized by membrane filtration or UV radiation. Apart from this plastic items, membrane filters, backelite screw caps, rubber fittings, Teflon wares, cotton wool, filter papers, etc., can bc sterilized by autoclaving. In emergency glasswares and metal items can also be sterilized in autoclave.

WATER SAMPLE COLLECTION

Water samples are most perfectly collected in sterilized stoppered glass bottles of heat resistant borosilicate glass. Good quality autoclave, non-toxic and non-leaching plastic bottles can also be used. Running water from main stream should be sampled aseptically avoiding manual contamination in case of piped supplies, hand pumps and streams. Chlorinated water should be collected in sampling bottles containing sodium thiosulfate. Each bottle should be labeled with necessary information such as site, source, date, time or any other remark. Water samples should be filled once without rinsing, leaving, 1/3rd or 1/4th airspace in bottles. Sample filled bottles should be transferred into ice box (below 10°C) and transported to laboratory. Samples should be analysed within 6 hrs. of collection but in no condition after 24 hrs., provided that sample was stored in refrigerator (4°C).

Preparation of Decimal (Ten-Fold) Dilution

For accurate enumeration of test organism in a contaminated water sample, serial decimal (ten-fold) dilutions of that sample is prepared in a suitable diluent. Usually 9 ml of diluent is dispensed in test tubes of appropriate size , cotton plug and autoclaved for sterlization. For preparation of serial decimal dilution 1 ml of test sample is mixed with 9 ml of diluent. Now 1ml of this mixture is transferrred in to next 9 ml of diluent. This process may be continued to make anticipitated range of serial decimal dilution of test sample.

A. PLATE COUNT METHOD

This method is generally used to enumerate total count of viable aerobic and facultative anaerobic bacteria in a certain volume of sample (usually 0.1-1.0 ml) or its dilution using appropriate agar (solidified) culture medium (15 + 5 ml) in a petridish (95 x 17 mm). The inoculated agar plates are incubated at required temperature such as 4-10°C, 20-25°C, 35-37°C or 42-45°C for 18, 24, 48 or 72 hours.

The colonies of bacteria grown on/in the agar plate are counted manually or with the help of any colony counter. Bacterial counts between 30-300 colonies per plate are statistically acceptable and reported as "colony forming units" (c.f.u.) per ml. Plates with no colonies are reported as < 1 c.f.u./ml. Plates with > 300 colonies can be counted after dividing the plate area in convenient number of squares, instead of reporting as "too numerous to count" (TNTC). Bacterial counts per ml are computed by multiplying with average number of colonies per plate by the reciprocal of dilution used. This method can be used in the following two ways.

1. POUR PLATE METHOD

In this method, usually 1 ml of test sample is poured with a sterile pipette into a sterile petridish (95 x 17 mm) followed by immediately pouring 10 ml of melted appropriate agar medium maintained between 45-50°C in water bath until used. Sample and medium are mixed by rotating gently in the covered plate without spoilage of medium. After solidification (about 10 minutes), plates are inverted and incubated for bacterial growth.

This pour plate counting which is also known as standard plate counting, has several disadvantages that limits recovery of maximum number of bacteria irrespective of the medium and the incubation period and temperature used. The tempered medium at 45-50°C may cause heat shock to stressed bacteria. Apart from this, the colonies grew under the agar surface cannot be picked fot further identification etc. Therefore, method is not much acceptable.

2. SPREAD PLATE METHOD

This method is more popular than the former one, for bacterial counting in any liquid sample. In this method appropriate agar medium plates are prepared by pouring 15 + 5ml of autoclaved melted agar medium. After the solidification, plates are dried at about 50 + 5°C for 30-45 minutes by spreading open seperately in a drying cabinet.

Usually 0.1-0.5 ml of liquid sample or its dilution is poured on agar surface of plate and inoculum is spread over immediately after pouring in rotating manner gently to dryness with a 'L'shape sterilized (with spirit and flame) and cooled round glass rod spreader. Plates are inverted and incubated for bacterial growth. Bacterial coloniesin each plate are counted as earlier. This method is very common for enumeration of bacteria in liquid, minced homogenized, filter washed, swab washed samples, etc. It is not much recommended method for routine analysis for indicator bacteria such as coliforms, faecal coliforms and faecal streptococci, because dilution cannot be anticipated according to the degree of contamination in test sample.

B. MEMBRANE FILTRATION METHOD

This technique is highly reproducible and can be used for relatively large volume of water samples. The standard volume to be filtered for drinking water is 100 ml. The method yields definite results more rapidly than multiple tube techniques, although it has limitations in testing turbid water without test bacteria. The technique is extremely useful in monitoring drinking water emergencies and analysis of a variety of natural waters.

The filtration unit is the main requirement of this method. It consists of water holder to hold non-toxic nitrocellulose-membrane filter of 0.2 or 0.45μ porosity and required diameter. Filtration unit and membrane are sterilized before each test. Receptacle of filtration unit is tightly mounted on the mouth of conical filtering flask of 1000ml volume with a side tube in neck region connected with an electric vacuum pump, water pressure pump, hand aspirator or any device which can exert pressure differential on membrane filter.

Usually 10-100 ml volumes of water filters is filtered through a sterile membrane filter. After complete filtration the membrane is transferred aseptically to agar medium or broth medium soaked (2 + 0.2 ml) non-toxic absorbent pad (consist of compressed filter paper disks of 47 x 0.8 mm or other material of high quality). Keeping the surface with trapped bacteria upward without air bubbles. Plates are inverted and incubated at 35 + 0.5°C for 24-48 hours. Plates with 20-200 colonies per membrane are usually considered for computation by the following equation:

$$\frac{\text{Total colonies / 100ml colonies}}{\text{ml sample filtered}} \times 100$$

If there is confluent growth and > 200 colonies per membrane, the result may be reported as 'confluent growth' or 'too numerous to count' (TNTC).

C. MULTIPLE TUBE FERMENTATION METHOD

This method is most acceptable and applicable using selective liquid culture medium to determine the degree of bacterial contamination as most probable number (MPN)/100 ml of test liquid or water dissolved sample or it's dilution. The method is based on the law of probability and multiple dilution to extinction. The results are reported as MPN/100 ml computed according to the Thomas formula as given below:

$$\text{MPN / 100 ml} = \frac{\text{No. of positive tubes}}{(\text{ml sample in negative tubes}) \times (\text{ml sample in all tubes})} \times 100$$

McCrady (1918) introduced probability table for statistical estimation of MPN/ 100 ml by adapting Thomas formula. This table is applicable to determine the MPN value for any higher dilution series, e.g., .01, .001, .0001 ml, not mentioned in the probability table and the positive tube combination is 5-3-0 respectively. The following formula is applicable:

$$\text{MPN /7100 ml} = \frac{\text{Table value}}{\text{Starting volume of sample}} \times 10$$

$$\text{For example} = \frac{80}{-01} \times 10$$

$$= 80{,}000$$

In this method, replicable fermentation tubes (usually 5 or 3) with double strength culture broth for 50 and 10 ml volumes, while single strength culture broth for 5 ml volume of medium with or without inverted vials (Durham's tube) are prepared. Autoclaved fermentation tubes are inoculated with same volume of test sample in 50 to 10 ml medium containing tubes, while others are inoculated with $1/10^{th}$ 0.1, 0.001.... and so on, upto end point, i.e., bacteria-free dilution. Inoculated sets of tubes are incubated usually at 37^0C (44^0C for faecal coliform and other thermotolerant bacteria) and observed after 24-48 hours for fermentation tubes with positive results.

The criteria for the positive result in fermentation tubes of presumptive test are the production of acid and/or gas after the fermentation of carbohydrates (usually sugars) essentially added in the broth (liquid) culture medium alongwith a suitable pH, sensitive stain and other ingredients. The acid production indicated by the colour change of medium and the produced gas is trapped in inverted vial (Durham's tube).

PREPARATION AND STERILIZATION OF CULTURE MEDIA, DILUENT AND REAGENT

A. MacConkey's Broth

Prepare 8% (w/v) double strength solution of MacConkey's broth and dispense 10 ml volume in test tubes of 20 x 150 mm size, insert an inverted Durham's tube (Fermentation tube) and cotton plug the mouth. Similarly, pepare 4% (w/v) single strength solution of MacConkey's broth and dispense 5 ml volume in test tubes of 17 x 150 mm size, insert an inverted Durham's tube and cotton plug the mouth.

B. Brilliant Green Bile (BGB) Broth and Peptone Water

Prepare 4% (w/v) solutions of BGB broth and 1.5% (w/v) solution of peptone water. Dispense 5 ml volume of each in the test tubes of 17 x 125 mm size, insert an inverted Durham's tube and cotton plug the mouth.

C. Azide Dextrose Broth

Prepare 7.0% (w/v) double strength solution of this broth, dispense 10 ml volume in test tubes of 20 x 120 mm size and cotton plug. Likewise, prepare 3.5% (w/v) single strength solution of this broth, dispense 5 ml volume in the test tubes of 17 x 125 mm size and cotton plug the mouth.

D. Bile Esculin Azide Agar

Dissolve Bile Esculin Azide Agar 5.7% (w/v) solution, respectively, in conical flask of suitable size by gentle heating. Cotton plug the mouth of flasks.

Pool the cotton plugged test tubes of assorted sizes containing culture medium, in the wire gauze baskets or metallic containers. Keep the baskets, containers and agar medium flasks into the autoclave for steam sterilization. After releasing the pressure open the autoclave, bring out the basket, containers and flasks.

Pour approx. 20 ml volume of the molten agar medium immediately into sterilized plastic or glass petridish of 100 x 15 mm size near the burner aseptically and cover the plates and keep at room temperature to solidify. The agar plates are dried in drying cabinet at 60°C for 30 minutes and stored in refrigerator for use.

E. Ringer's Solution (Diluent)

Dissolve 5 gm of sodium chloride, 4 gm of potassium chloride and 3 gm of potassium dihydrogen phosphate in one litre of distilled water. Dispense the required volumes into dilution test tubes, bottles and flasks and cotton plug. Keep the test tubes and flasks with diluent in containers, put into the autoclave to sterilize.

F. Kovac's Indole Reagent

Dissolve 5 gm of p-dimethyl aminobenzaldehyde in 75 ml amyl alcohol by gently warming in water bath at 50-55°C temperature. Cool and add 25 ml of concentrated hydrochloric acid slowly. Protect from light and store in a dark colored glass bottle at 4°C.

ENUMERATION OF COLIFORMS AND FAECAL COLIFORMS

The test water sample is vigorously shaked (approx. 20 times) to make the sample homogenous. 10ml of samples is inoculated with a sterilized 10 ml pipettes in 5 tubes (1 set) containing 10 ml of double strength MacConkey's broth with a inverted Durham's tube in each. Then 1 ml sample is transferred with 1 ml sterilized pipette in 5 tubes (II set) containing 5 ml of single strength MacConkey's broth with an inverted Durham's tube in each. This process of transferring 1ml of sample is repeated with 1/10th dilution of sample for III set of 5 ml single strength MacConkey's broth with an inverted Durham's tube in each. To enumerate the coliform in highly contaminated water samples, next serial higher dilutions may be similarly inoculated.

All inoculated tubes are incubated at 37°C. After 24-48 hrs number of positive tubes showing (color change from red to orange/yellow) and gas production (trapped in inverted Durham's tube) are counted and MPN of PRESUMPTIVE COLIFORM/100 ml of sample is worked out by consulting probability. To determine MPN of CONFIRMED COLIFORM/100 sample, all positive tubes are presumptive coliform test are subcultured with inoculation loop (Nichrome wire) in test tubes containing 5 ml of single strength MacConkey's broth with an inverted Durham's tube in each and incubated at 37°C. After 48 hrs number of positive tubes showing acid and gas production confirm the presence of coliform.

This test can be minimized by using either 5 tubes with 10 ml of triple strength of medium or 10 tubes with 10 ml of double strength of medium inoculated with 20 ml or 10 ml of sample; respectively MPN of coliform/100 ml is determined with number tubes appeared positive after inocubation at 37°C for 48 hrs., by consulting probability.

For enumeration of faecal coliform, all the positive tubes of confirmed coliform test are subcultured with inoculation loop from each positive tubes in to 5 ml of Brilliant Green Bile (BGB) broth with an inverted Durham's tube and 5 ml of Peptone water tubes. The subcultured tubes are incubated at 44°C (preferably in water bath). After 24-48 hrs., BGB broth tubes with growth and gas production are counted to determine the MPN of PRESUMPTIVE FAECAL COLIFORM/100 ml of sample. 4-5 drops of Kovac's reagent is added with pipette in Peptone water tubes paired with positive BGB broth tubes. The positive Peptone water tubes showing red layer on surface are counted to determine MPN of CONFIRMED FAECAL COLIFORM 100 ml of sample.

ENUMERATION OF FAECAL STREPTOCOCCI

Like coliform test, 3 sets of 5 tubes, i.e., one set of 10 ml double strength and two sets of 5 ml single strength Azide Dextrose broth are inoculated with 10 ml, 1 ml and 0.1 ml of water sample, respectively. Inoculated tubes are incubated for 48 hours at 37°C. Positive tubes showing bacterial growth with or without color change from purple to straw yellow, are streaked separately on the Bile Esculin Azide agar plates and incubated at 44°C. After 24 hours, plates with brownish black cloudy colonies are counted and MPN of FAECAL STREPTOCOCCI is determined byconsulting the probability table.

Multiple tube fermentation method is commonly used for bacteriological analysis of drinking water, soil, sediment, sludge, slurry, etc. to determine the MPN of pollution indicator bacteria such as , coliform, faecal coliform and faecal streptococci including *Clostridium* sp., *Aeromonas* sp., *Vibrio* sp., *Pseudomonas* sp. etc., followed by the recommended specific confirmatory tests.

This method has been simplified using single set of 10 tubes with 10 ml. of double strength selective culture broth medium for enumeration of coliform bacteria specially drinking water. This method provides counts upto > 23 bacteria/100 ml. Therefore, this simplification is suitable for the quantitative analysis of finished water samples, particularly with large number of samples in laboratory or in field as well as during emergencies and outbreaks.

According to Bureau of Indian Standard Drinking Water Specifications (BIS, 2003) and Guidelines for Drinking Water Quality (WHO, 2004), the drinking water should be with < 10 coliforms/100 ml and no faecal coliform, faecal streptococci or any other pathogen.

PRESENCE- ABSENCE (P-A) TEST

This is one tube or bottle method for screening the bacteriological quality for indicator bacteria in large number of samples. In this test one large volume (50-100 ml) of test sample is inoculated in 50-100 ml double strength broth culture medium suitable for the test bacteria P-A culture bottle (250 ml. screw- capped milk dilution bottle) containing a sterile inverted vial. Inoculated bottles are inocubated at recommended temperature (usually 37°C). After 24-48 hours, bottles are inspected for colour change due to acid production (and also gas collection in the inverted vial in case of coliform) to determine the presence or absence of test bacteria.

The test provides qualitative results for coliform or any other bacteria such as faecal coliform, faecal streptococci, *Aeromonas* sp., *Staphylococcus* sp., *Shigella* sp., *Clostridium* sp. etc. in terms of 'Present' or 'Absent' in test volume of sample. Thus, for the qualitative screening of a small or large number of samples, P-A test is convenient, economical and time saving.

COMPOSITION OF CULTURE MEDIUM USED

Medium	Component	Quantity
1. MacConkey's broth	Peptic digest	20.00
	Lactose	10.00
	Bile salts	05.00
	Sodium chloride	05.00
	Neutral red	0.075
2. Brilliant Green Bile broth	Peptic digest	10.00
	Lactose	10.00
	Ox gall	20.00
	Brilliant green	0.0133
3. Peptone water	Peptone	10.00
	Sodium chloride	05.00
4. Nutrient agar (1.5%)	Beef extract	03.00
	Peptone	05.00
	Sodium chloride	08.00
	Agar	15.00
5. Azide dextrose broth	Peptone	15.00
	Beef extract	04.5
	Dextrose	07.50
	Sodium chloride	07.50
	Sodium azide	00.20
6. Bile esculin azide agar	Beef extract	03.00
	Peptone	05.00
	Ox gall	40.00
	Esculin	01.00
	Ferric citrate	0.50
	Agar	15.00nm

QUALITATIVE AND QUANTITATIVE ANALYSIS OF PLANKTONS

The planktons are the heterogenous assemblage of minute organisms which occur in natural water and float about by the wave action and movement of water. Plankton form the integral part of any aquatic system and include phytoplankton and zooplankton as free-floating autotrophs. Their density and the species diversity indicates the quality of water in natural system. Phytoplankton are microscopic free-floating autotrophs. These phytoplankton are distributed with respect to different seasons and pollution load. Phytoplankton are represented by four groups of algae (i) Bacillariophyceae (yellow green diatoms), (ii) Chlorophyceae (green algae), (iii) Cynophyceae (blue-green algae), (iv) Euglenophyceae and dinoflagelates. When they are abundant, they may cause the water to look cloudy or may float on the surface as 'scum'. Such heavy growth is called as 'bloom'. Zooplankton are minute organism that float on the water surface along with water current. They may bear some locomotary structures but are not capable of moving against water current. They usually act as primary consumers and constitute an important link between primary producers (Phytoplankton) and higher consumers (like carnivorous fishes) and aquatic chain. They occupy an intermediate position in food web. Many of them feed on bacteria and algae and in turn are fed by vertebrate, fishes and birds. The seasonal changes in zooplankton species are clearly related to the physio-chemical and biological regime of the aquatic environments. Freshwater zooplanktons are represented by protozoans, crustaceans, rotifers copepoda, cladocera and obstracoda.

Plankton being sensitive to temperature, dissolve oxygen, solar radiation and nutrient level of water, show vertical movements. The optimum conditions prevail during morning hours. Therefore sampling should be done between 7.00 a.m. to 9.00 a.m. at different depth in the surface, at 10 feet and 25 feet. Plankton samples are conveniently collected using the following devices:

Collection and Preservation of Plankton

Zooplankton and phytoplankton may be collected by filtering a known amount of water through a plankton net made by bolting silk (no. 25, i.e., mesh size 0.55 ìm) for phytoplankton. Surface water may be collected with the help of plastic bucket of known volume but, for sub-surface water, water samples are considered to be better device.

Plankton Net

When over all phytoplankton and zooplankton composition and population, irrespective of depth to be analysed various types of plankton nets are used but a simple conical net is popularly used for many years. It is conical in shape and made with standard silk bolting cloth number 22 having 75 meshes/linear centimeter. The diameter top metal ring is 36 centimeters to which is attatched wooden handle, length of cone is 36 cms; at the bottom of which graduated 10 ml vial is tied firmly.

Immerse the ring of the net completely below the water surface to a desired depth. Haul the net against the water current to a desired distance (1) Transfer the contents of the specimen tube to another tube without loss of samples. Repeat the hauling five times in the above manner to collect replicates. Shake the sample well and immediately take out 5.0 ml sample into another tube for plankton counting. Preserve other portion in 4 percent formaline for zooplanktons and in Lugal's iodine solution for phytoplankton identification. Calculate the volume of the water filtered through net using the following formula :

$V = r^2 l$

where, V = volume of water filtered

R = radius of plankton net ring

l = column of water filtered

Water Sampler

Water sampler is used for collection of the sample of vertical column. Sampler has a wide mouth, rubber stopper bottle (2 to 5 litre capacity), covered with an iron sheet or zinc sheet container. The space between the bottle and the container is filled with sawdust. The rubber stopper has a thermometer, and an inlet and an outlet tube. Outlet tube is attatched to a long plastic tube with a stopper clip at its free end. The container is placed in a strong nylon netted bag provided at its base with a weight. The nylon bag is tied to a rope marked in meters. Because these samples collect whole water samples, all size classes of phytoplankton are collected.

Sampling Procedure

The sampler is dipped in water body to desired depth. The stopper clip is pressed by the operator, the water rushes inside the bottle to fill the air space. The clip stopper is closed and sampler is lifted up. The temperature of water at particular depth is recorded. Remove the rubber stopper and filter the water for plankton. Because these sampler collect whole water samples, all size classes of plankton are collected. Different size categories of plankton can be separated by subsequently from these whole water samples through netting by the appropriate mesh size. Select appropriate mesh sizes for concentrating the various size categories of plankton typical of the aquatic system under study. For collection of micro zooplankton such as protozoans, rotifers and immature crustaceans 20 to 200μm net is used. The larger and mature crustaceans may be concentrated by passing through a 20μm mesh net. (APHA, 1998).

Plankton Counting

To enumerate plankton use a counting cell chamber that limits the volume and area for ready calculation. Plankton can be counted by the following two methods. The 10 ml sample collected in graduated tube is subjected to centrifugation with the help of centrifuge at 3000 R.P.M. for 20 minutes. Supernatant fluid is decanted and concentrate is thoroughly mixed and used for enumeration.

Simplified Method:

Shake the sample thoroughly. Out of unfixed 5.0 ml sample take out 1.0 ml sample in caliberated pipette (attatched with a sucker at sucking end). Place the sample drop by drop (each drop separately) over a glass slide and cover it with cover glass. Count the number of the total organisms (zoo and phytoplanktons separately) in each drop in 1.0 ml of the sample. Sum up the number of organisms of all the drops. The number will represent the total organisms in 1.0 ml of sample.

Formula used for calculation of Phytoplankton as units/L is :

$$\text{Plankton density unit / L} = \frac{n \times c}{V} \times 1000$$

Where

n = no. of plankton counted in 1.0 ml concentrate

c = total volume of concentrate in ml

V = total volume of water filtered through net

Identify the plankton with the help of standard keys and books by Palmer (1980) and APHA (2005).

Microtransect Method:

Take out 0.1 ml of the concentrate plankton sample (after thorough mixing) over a clean slide by a dropper. Cover with 22 mm x 22 mm cover slip. Count the organisms in 5 strips each having 6 different places. Each visible area (focus) will represent one microtransect. Find out the area of the transectand width of high power of microscope field. Count quickly to avoid drying of the drop. Repeat the process with atleast five drops (Lackey, 1938) and subsequently modified by (Edmond, 1974). Identification of phytoplankton can be done upto genera level with the help of standard keys and books by Palmer (1980), APHA 2005.

$$\text{Plankton density (number / L)} = \frac{a^2 \times v^1}{V}$$

where,

Total number of organisms per drop (0.1 ml) a^1

Number of planktons per ml.

a^2 = number of plankton per ml

V^1 = Volume of concentrate collected in specimen tube

V = Volume of total water filtered through plankton net

COUNTING BY SEDWICK RAFTER CELL (S-R)

The Sedgwick cell is a device commonly used for plankton counting. Approximately 50 mm long x 20 mm wide x 1 mm deep. The total area of the bottom is approximately 1000 mm^2 and total volume is 1000 mm^2 are 1.0 ml.

Before filling the S.R. cell with sample, place the cover slip diagonally across the cell and transfer sample with pipette. Placing the cover slip in this manner will be helpful to prevent formation of bubbles in the corner. Before counting, let the S.R. cell stand atleast for 15 minutes to settle the plankton. Count the plankton on the bottom of S.R. cell with the help of keys and books under inverted compound microscope at 100X-200X magnification for phytoplankton. Using compound microscope and magnification of 100X enumerate zooplankton (protozoa, totifire and naupli). A strip is about 50 mm long, 1 mm deep and width 0.5 mm (500 urn) (Palmar 1969, APHA 2005). Derive number of planktons in the S.R. cell from the following :

$$No / ml = \frac{C \times 1000 \text{ mm}^2}{L X D X W X S}$$

where: C = number of organism counted

L = length of each strip

D = depth of strip

S = number of strips counted

COUNTING BY HEMOCYTOMETER

Another readily available chamber is standard medical hemocytometer used for enumerating blood cells. It has ruled grid machined into a counting plate in fitted with a ground-glass coverslip. The grid is divided into 1.0 mm sq. division, the chamber is 0.1 mm deep. Introduce the sample below the cover slip by pipette and view under 450X magnification. Keep the slide undisturbed for 1-2 minute to settle the suspension. Cells are counted in 4 large squares chosen randomly. The large square measure 1 mm along with slide. Therefore, the entire grid is 16 mm sq. Hemocytometer is 0.2 mm deep. Therefore, value of sample that cover grid is of 9 large square 0.009 ml. The chamber comes from the manufacturer with detailed instructions sheet containing direction, calculation and proper usage (APHA, 2005).

ENUMERATION OF ZOOPLANKTON (SMALL)

An open counting chamber 80 by 50 mm and 2 mm deep, with circular grooves or partition of 1-5 ml capacity for small zooplankton (protozoans, rotifers and nuplii) and 5 to 10 ml for larger, mature microcrustaceans is used as counting chamber. A mild detergent solution is placed on the chamber before counting to reduce movement of organisms. Counting of the organism is done by using compound microscope and a magnification of 100X. A Sedgwick-Rafter cell is not suitable because of its size. Smaller zooplankton as number per litre and larger forms as number per cubic meter should be reported:

$$No. / m^3 = \frac{C X V^2}{V^1 X V^{11}}$$

where c = number of organism

V^1 = value of the concentrated sample

V^{11}= volume counted, ml, and

To obtain organisms per litre divided by 1000.

Diversity Indices

Species diversity is a statistical abstraction with two components reflecting the number of species (richness) and distribution of individual species (richness) and distribution of individual of all species at a particular site. The two types of diversity indices are computed. Following diversity indices are commonly used.

Odum's species per thousand individual

Odum et al. (1960) proposed that the species per thousand individuals can be used as an index of diversity which is a basically an evenness and is very reliable index. It is calculated as:

$$\text{Odum's index} = \frac{\text{No. of total species in the sample}}{\text{No. of total individuals of all species}} \times 1000$$

REFERENCES

APHA (2005). Standard Methods for the Examination of Water and Wastewater. 21st Edn. APHA, AWWA and WPCF, Washington.

BIS (2003) Indian Standard: Drinking Water Specification. First Revision IS 1500: 1991 Bureau of Indian Standards, New Delhi.

McCrady, MH (1918) Tables for rapid interpretation of fermentation tube results. The Public Health Journal, Canada, 9,201.

Odum, H.T. Cantlon J.E. and Kornickar, L.S. (1960). An organisational hierarchy postulate for interpretation of species individuals, distribution, species entrophy and ecosystem evaluation and the meaning of a species variety index. Ecology 41, 395-399.

Palmar, C.M.(1980). Algae and Water Pollution. Castle House Publication Ltd. 1-119.

WHO (2004) Guidelines for drinking water 3rd edition. World Health Organization, Geneva.

Chapter 9

Introduction to Bioinorganic Chemistry

Bioinorganic chemistry is a specialized field that spans the chemistry of metal-containing molecules within biological systems. This field is concerned with the control and use of metal ions in biochemical processes. Although bioinorganic chemistry includes the study of artificially introduced metals (e.g. medicinally), many natural occurring biological processes (such as respiration) depend upon molecules containing inorganic elements, such as metalloproteins, and these natural processes are also studied by bioinorganic chemistry. Bioinorganic chemistry has developed from the continuing research in inorganic chemistry and its important associations in biological chemistry.

Bioinorganic chemistry is that branch, which deals with the study of metal-containing molecules within biological systems.

It is a discipline which is focused on:

- the study of elements and inorganic compounds in biological systems and the study of inorganic models that could imitate the chemical behavior in those biological systems.
- the study of non-essential elements with an application in the diagnosis and therapeutic field.

As a mix of biochemistry and inorganic chemistry, bioinorganic chemistry is important in realizing the implications of electron-transfer proteins, substrate bindings and activation, atom and group transfer chemistry as well as metal properties in biological chemistry.

Metal ion transport and storage covers a diverse collection of ion channels, ion pumps (e.g. NaKATPase), vacuoles, siderophores, and other proteins and small molecules whose aim is to carefully control the concentration of metal ions in the cell (sometimes referred to as metallome).

Hydrolase enzymes include a diverse collection of proteins that interact with water and substrates. Examples of this class of metalloproteins are carbonic anhydrase, metallophosphatases, and metalloproteinases.

Metal containing electron-transfer proteins are organized into three major classes:

- iron-sulfur proteins such as rubredoxins, ferredoxins, and Rieske proteins
- blue copper proteins
- cytochromes.

These electron transport proteins are complementary to the non-metal electron transporters i.e. nicotinamide adenine dinucleotide (NAD) and flavin adenine dinucleotide (FAD).

Oxygen transport and activation proteins make extensive use of metals such as iron, copper, and manganese. Heme is utilized by red blood cells in the form of hemoglobin for oxygen transport and is perhaps the most recognized metal system in biology. Other oxygen transport systems include myoglobin, hemocyanin, and hemerythrin. Oxidases and oxygenases are metal systems found throughout nature that take advantage of oxygen to carry out important reactions such as energy generation in cytochrome c oxidase or small molecule oxidation in cytochrome P450 oxidases or methane monooxygenase. Some metalloproteins are designed to protect a biological system from the potentially harmful effects of oxygen and other reactive oxygen-containing molecules such as hydrogen peroxide. These systems include peroxidases, catalases, and superoxide dismutases. A complementary metalloprotein to those that react with oxygen is the oxygen evolving complex present in plants. This system is part of the complex protein machinery that produces oxygen as plants perform photosynthesis.

Bioorganometallic systems such as hydrogenases and methylcobalamin are biological examples of organometallic chemistry.

The nitrogen metabolism pathways make extensive use of metals. Nitrogenase is one of the more famous metalloproteins associated with nitrogen metabolism. More recently, the cardiovascular and neuronal importance of nitric oxide has been examined, including the enzyme nitric oxide synthase.

Metals in medicine is the study of the design and mechanism of action of metal-containing pharmaceuticals, and compounds that interact with endogenous metal ions in enzyme active sites. This diverse field includes the platinum and ruthenium anti-cancer drugs, chelating agents, gold drug chaperones, and gadolinium contrast agents.

Trace Element

Early in the twentieth century, scientists were able to qualitatively detect small amounts of several elements in living organisms. At the time, these elements were often described as being present in "traces" or "trace amounts." This apparently led to the term "trace elements", which today is usually defined as mineral elements that occur in living systems in micrograms per gram of body weight or less. A majority of elements of the periodic table probably could be considered trace elements. However, the presence of most of these elements in higher animals quite likely is just a manifestation of our geochemical origin or the result of environmental exposure. Only eight trace elements are generally accepted as being essential for health and well-being in higher animals through the consumption of food and beverages; these are cobalt, copper, iodine, iron, manganese, molybdenum, selenium, and zinc. Evidence has recently appeared that indicates two other trace elements, boron and chromium, may also be essential; however, general acceptance of their essentiality is still lacking. Based on findings with experimental animals and lower-life forms, numerous other trace elements have been suggested as being essential for higher animals including aluminum, arsenic, fluorine, lithium, nickel, silicon, and vanadium. However, conclusive evidence for essentiality, such as a defined biochemical function, is lacking for these elements. Thus, their nutritional importance remains to be determined. Of the elements mentioned, assuring the consumption of foods providing adequate amounts of iodine, iron, and zinc is of greatest practical concern for human health. Evidence is emerging, however, suggesting that the amount of cobalt (as vitamin B_{12}), copper, selenium, boron, and chromium provided through foods should be considered a practical nutritional concern in assuring or promoting health and well-being.

Physiological Roles of Trace Elements

Trace elements have several roles in living organisms. Some are essential components of enzymes where they attract substrate molecules and facilitate their conversion to specific end products. Some donate or accept electrons in reactions of reduction and oxidation, which results in the generation and utilization of metabolic energy. One trace element, iron, is involved in the binding, transporting, and releasing of oxygen in higher animals. Some trace elements impart structural stability to important biological molecules. Finally, some trace elements control important biological processes through such actions as facilitating the binding of molecules to receptor sites on cell membranes, altering the structure or ionic nature of membranes to prevent or allow specific molecules to enter or leave a cell, and inducing gene expression resulting in the formation of proteins involved in life processes.

Homeostatic Regulation of Trace Elements

The ability of the body to maintain the content of a specific substance such as a trace element within a certain range despite varying intakes is called homeostasis. Homeostasis involves the processes of absorption, storage, and excretion. The relative importance of these three processes varies among the trace elements. The homeostatic regulation of trace elements existing as positively charged cations (for example, copper, iron, zinc) occurs primarily during absorption from the gastrointestinal tract. Trace elements absorbed as negatively charged anions (for example, boron, selenium) are usually absorbed freely and completely from the gastrointestinal tract. Thus, they are homeostatically regulated primarily by excretion through the urine, bile, sweat, and breath. Storage of trace elements in inactive sites or forms is another mechanism that prevents inappropriate amounts of reactive trace elements to be present, for example, storage of iron in the form of ferritin. Release of a trace element from a storage site also can be important in preventing deficiency.

Iodine

This trace element has one known function in higher animals and humans. It is a constituent of thyroid hormone (thyroxine, T_4) that, after conversion to triiodothyronine (T_3), functions as a regulator of growth and development by reacting with cell receptors, which results in energy (adenosine triphosphate, ATP) production and the activation or inhibition of synthesis of specific proteins.

Recognition that iodine was nutritionally important began in the 1920s when it was found that iodine prevented goiter, and increased iodine intake was associated with decreased endemic cretinism, the arrested physical and mental development caused by the lack of thyroid hormone. Today, the consequences of iodine deficiency still are a major public health problem in the world. In fact, iodine deficiency is the most prevalent global cause of preventable mental retardation. Briefly, the spectrum of iodine deficiency disorders is large and includes fetal congenital anomalies and perinatal mortality; neurological cretinism characterized by mental deficiency, deaf mutism, spastic diplegia (spastic stiffness of the limbs), and squint; psychomotor defects; goiter; and slowing of the metabolic rate causing fatigue, slowing of bodily and mental functions, weight increase, and cold intolerance.

Although homeostatic mechanisms allow for a substantial tolerance to high intakes of iodine, iodine-induced hyperthyroidism has been recognized for nearly two centuries. People who have had a marked iodine deficiency and are then given high amounts of iodine as part of a preventative program are at risk of getting hyperthyroidism with clinical signs including weight loss, tachycardia, muscle weakness, and skin warmth.

Iodide, an anion, is rapidly and almost completely absorbed from the stomach and upper gastrointestinal (GI) tract. Most other forms of iodine are changed in the GI tract to iodide and completely absorbed. When thyroid hormone is ingested, about 80 percent is absorbed without change; the rest is excreted in the faeces. Absorbed iodide circulates in the free form; it does not bind to proteins in blood. Iodide is rapidly removed from circulation by the thyroid and kidney. Urinary excretion is a major homeostatic mechanism. If iodine intake has been adequate, only about 10 percent of absorbed iodide appears in the thyroid, the rest appears in the urine. However, if iodine status is inadequate, a much higher percentage, upto 80 percent, can appear in the thyroid. The thyroid gland is essentially the only storage site for iodine, where it appears mostly as mono- and diiodotyrosine and T_4, with a small amount of T_3.

Iron

This trace element is a component of molecules that transport oxygen in blood. Numerous enzymatic reactions involving oxidation and reduction (redox) use iron as the agent through which oxygen is added, hydrogen is removed, or electrons are transferred. The classes of enzymes dependent on iron for activity include the oxidoreductases, examplified by xanthine oxidase/dehydrogenase; monooxygenases, examplified by the amino acid oxidases and cytochrome P450; dioxygenases, examplified by amino acid dioxygenases, lipoxygenases, peroxidases, fatty acid desaturases, and nitric oxide synthases; and miscellaneous enzymes such as aconitase.

Among the trace elements, iron has the longest and best described history. By the seventeenth century, a recognized treatment for chlorosis, or iron deficiency anemia, was drinking wine containing iron filings. Despite the extensive knowledge about its treatment and prevention, and the institution of a variety of effective interventions, iron deficiency is the primary mineral deficiency in the United States and the World today. The physiological signs of iron deficiency include anemia, glossitis (smooth atrophy of tongue surface), angular stomatitis (fissures at the angles of the mouth), koilonychia (spoon nails), blue sclera, lethargy, apathy, restlessness, and fatigue. Pathological consequences of iron deficiency include impaired thermoregulation, immune function, mental function, and physical performance; complications in pregnancy, including increased risk of premature delivery, low birth weight, and infant mortality; and possibly increased risk of osteoporosis.

Concerns have been expressed about high intakes of iron being a health issue. This has come about through epidemiologic observations associating high dietary iron or high body iron stores with cancer and coronary heart disease. Further experimental studies, however, are required to confirm whether the high intakes of iron increase the risk for these diseases. The toxic potential of iron arises from its biological importance as a redox element that accepts and donates electrons to oxygen that can result in the formation of reactive oxygen species or radicals that can damage cellular components such as fatty acids, proteins, and nucleic acids. Antioxidants are enzymes or molecules that prevent the formation of oxygen radicals or convert them to non-radical products. When not properly controlled by antioxidants, reactive oxygen damage can lead to premature cell aging and death. An iron overload disease known as hereditary hemochromatosis is caused by a defective regulation of iron transport with excessive iron absorption and high transferrin (transport form) iron in plasma. Clinical signs appear when body iron accumulates to about 10 times normal and include cirrhosis, diabetes, heart failure, arthritis, and sexual dysfunction. Hemochromatosis also increases the risk for hepatic carcinoma. The treatment for hereditary hemochromatosis is repeated phlebotomy.

Absorption from the GI tract is the primary homeostatic mechanism for iron. Dietary iron exists generally in two forms, heme and non-heme, that are absorbed by different mechanisms. Heme iron is a protoporphyrin molecule containing an atom of iron; it comes primarily from hemoglobin and myoglobin in meat, poultry, and fish. Non-heme iron is primarily inorganic iron salts provided mainly by plant-based foods, dairy products, and iron-fortified foods. Heme iron is much better absorbed and less affected by enhancers and inhibitors of absorption than non-heme iron. Iron absorption is regulated by mucosal cells of the small intestine, but the exact mechanisms in the regulation have not been established. Both iron stores and blood hemoglobin status have a major influence on the amount of dietary iron that is absorbed. Under normal conditions, men absorb about 6 percent and menstruating women absorb about 13 percent of dietary iron. However, with severe iron deficiency anemia (functionally deficient blood low in hemoglobin), absorption of non-heme iron can be as high as 50 percent. Iron loss from the body is very low, about 0.6 mg/day. This loss is primarily by excretion in the bile and, along with iron in desquamated mucosal cells, eliminated via the faeces. Menstruation is a significant means through which iron is lost for women. However, that non-physiological loss of iron resulting from conditions such as parasitism, diarrhea, and enteritis account for half of iron deficiency anemia globally. Excess iron in the body is stored as ferritin and hemosiderin in the liver, reticuloendothelial cells, and bone marrow.

The recommended intakes of iron for the RDA for adult males and post-menopausal women is 8 mg/day; and that for menstruating adult women is 18 mg/day. Meat is the best source of iron, but iron-fortified foods (cereals and wheat-flour products) also are significant sources.

Zinc

This trace element is the only one that is found as an essential component in enzymes from all six enzyme classes. Over 50 zinc metalloenzymes have been identified. Zinc also functions as a component of transcription factors known as zinc fingers that bind to DNA and activate the transcription of a message, and imparts stability to cell membranes.

Signs of zinc deficiency in humans were first described in the 1960s. Although it is generally thought that zinc deficiency is a significant public health concern, the extent of the problem is unclear because there is no well-established method to accurately assess the zinc status of an individual. The physiological signs of zinc deficiency include depressed growth; anorexia (loss of appetite); parakeratotic skin lesions; diarrhea; and impaired testicular development, immune function, and cognitive function. Pathological consequences of zinc deficiency include dwarfism, delayed puberty, failure to thrive (acrodermatitis enteropathica infants), impaired wound healing, and increased susceptibility to infectious disease. It has also been suggested that low zinc status increases the susceptibility to osteoporosis and to pathological changes caused by the presence of excessive reactive oxygen species or free radicals.

Zinc is a relatively non-toxic element. Excessive intakes of zinc occur only with the inappropriate intake of supplements. The major undesirable effect is an interference with copper metabolism that could lead to copper deficiency. Long-term high zinc supplementation can reduce immune function and high-density lipoprotein (HDL)-cholesterol (the "good" cholesterol). These effects are seen only with zinc intakes of 100 mg/day or more.

A primary homeostatic mechanism for zinc is absorption from the small intestine. Absorption involves a carrier-mediated component and a non-mediated diffusion component. With normal dietary intakes, zinc is absorbed mainly by the carrier-mediated mechanism. Although absorption can be modified by a number of factors, about 30 percent of dietary zinc is absorbed. The efficiency of zinc absorption is increased with low zinc intakes. The small intestine has an additional role in zinc homeostasis through regulating excretion through pancreatic and intestinal secretions. After a meal, greater than 50 percent of the zinc in the intestinal lumen is from endogenous zinc secretion. Thus, zinc homeostasis depends upon the reabsorption of a significant portion of this endogenous zinc. Intestinal conservation of endogenous zinc apparently is a major mechanism for maintaining zinc status when dietary zinc is inadequate. The urinary loss of zinc is low and generally not markedly affected by zinc intake.

The recommended intakes of zinc for RDA: in adult males it is 11 mg/day and for adult females is 8 mg/day. The best food sources for zinc are red meats, organ meats (for example, liver), shellfish, nuts, whole grains, and legumes. Many breakfast cereals are fortified with zinc.

Cobalt (Vitamin B_{12})

Ionic cobalt is not an essential nutrient for humans. Cobalt is an integral component of vitamin B_{12}, which is an essential nutrient for non-ruminant animals and humans. Vitamin B_{12} is a co-factor for two enzymes, methionine synthase which methylates homocysteine to form methionine, and methylmalonyl coenzyme A (CoA) mutase which converts L-methylmalonyl CoA, formed by the oxidation of odd-chain fatty acids, to succinyl CoA.

In the nineteenth century, a megaloblastic anemia (functionally deficient blood containing primitive large red blood cells) was identified that was invariably fatal and thus called pernicious anemia. The first effective treatment for this disease was 1 pound of raw liver daily. In 1948, the pernicious anemia factor (vitamin B_{12}) in liver was isolated and found to contain 4 percent cobalt. Vitamin B_{12} deficiency most commonly arises when there is a defect in vitamin B_{12} absorption caused by such factors as atrophic gastritis, *Helicobacter pylori* infection, and bacteria overgrowth resulting from achlorohydria and intestinal blind loops. As vitamin B_{12} only comes from foods of animal origin, absolute vegetarianism will lead to deficiency in vitamin B_{12} after 5 to 10 years. The physiological signs of severe vitamin B_{12} deficiency are megaloblastic anemia, spinal cord demyelination, and peripheral neuropathy. The pathological consequences of deficiency include pernicious anemia, memory loss, dementia, an irreversible neurological disease called subacute degeneration of the spinal cord, and death. Recently, mild vitamin B_{12} deficiency has been cited as a cause of high circulating homocysteine, which has been associated with an increased risk for cardiovascular disease. Vitamin B_{12} is essentially non-toxic. Doses upto 10,000 times, the minimal daily adult human requirement do not have adverse effects.

Vitamin B_{12} absorption is a relatively complex process. Digestion by the saliva and acid environment of the stomach releases vitamin B_{12} from food, then it is bound to a haptocorrin called R protein that carries it into the duodenum. A binding protein, called intrinsic factor, released by gastric parietal cells binds vitamin B_{12} after the stomach acid is neutralized in the duodenum, and digestive enzymes remove the R binder from the vitamin. The intrinsic factor bound vitamin B_{12} is carried to a specific receptor in the ileum called cubilin and internalized by receptor-mediated endocytosis. Because vitamin B_{12} is water-soluble, excessive intakes are efficiently excreted in the urine.

The recommended intakes for vitamin B_{12} for various age and sex groups are shown in Table 1, which shows that the RDA for adults is 2.4 >g/day. Food sources of vitamin B_{12} are of animal origin and include meats, dairy products, and eggs. Fortified cereals have also become a significant source of vitamin B_{12}.

Copper

Copper is a co-factor for a number of oxidase enzymes including lysyl oxidase, ferroxidase (ceruloplasmin), dopamine beta-monooxygenase, tyrosinase, alpha-amidating monooxygenase, cytochrome C oxidase, and superoxide dismutase. These enzymes are involved in the stabilization of matrixes of connective tissue, oxidation of ferrous iron, synthesis of neurotransmitters, bestowal of pigment to hair and skin, assurance of immune system competence, generation of oxidative energy, and protection from reactive oxygen species. Copper also regulates the expression of some genes.

Although copper is a well-established essential trace element, its practical nutritional importance is a subject of debate. Well-established pathological consequences of copper deprivation in humans have been described primarily for premature and malnourished infants and include a hypochromic, normocytic, or macrocytic anemia; bone abnormalities resembling scurvy by showing osteoporosis, fractures of the long bones and ribs, epiphyseal separation, and fraying and cupping of the metaphyses with spur formation; increased incidence of infections; and poor growth. The consequences of the genetic disorder Menkes' disease (copper deficiency caused by a cellular defect in copper transport) in children include "kinky-type" steely hair, progressive neurological disorder, and death. Other consequences have been suggested based upon findings from epidemiological studies, and animal and short-term human copper deprivation experiments; these include impaired brain development and teratogenesis for the fetus and children, and osteoporosis, ischemic heart disease, cancer, increased susceptibility to infections, and accelerated aging for adults.

Copper toxicity is not a major health issue. The ingestion of fluids and foods contaminated with high amounts of copper can cause nausea. Because their biliary excretion pathway is immature, accumulation of toxic amounts of copper in the liver could be a risk for infants if intake is chronically high; this apparently caused cases of childhood liver cirrhosis in India.

Intestinal absorption is a primary homeostatic mechanism for copper. Copper enters epithelial cells of the small intestine by a facilitated process that involves specific copper transporters, or non-specific divalent metal ion transporters located on the brush-border surface. Then the copper is transported to the portal circulation where it is taken up by the liver and resecreted in plasma bound to ceruloplasmin. Transport of copper from the liver into the bile is the primary route for excretion of endogenous copper. Copper is of biliary origin and non-absorbed dietary copper are eliminated from the body via the faeces. Only an extremely small amount of copper is excreted in the urine. The absorption and retention of copper varies with dietary intake and status. For example, the percentages of ingested copper absorbed were 56 percent, 36 percent, and 12 percent with dietary intakes of 0.8, 1.7, and 7.5 mg/day, respectively. Moreover, tissue retention of copper is markedly increased when copper intake is low.

The recommended intakes for copper for various ages and sex groups, shows that the RDA for adults is 900 >g/day. The best sources of copper are legumes, whole grains, nuts, organ meats (for example, liver), seafood (for example, oysters, crab), peanut butter, chocolate, mushrooms, and ready-to-eat cereals.

Selenium

Selenium is a component of enzymes that catalyze redox reactions; these enzymes include various forms of glutathione peroxidase, iodothyronine 5<-deiodinase, and thioredoxin reductase.

Although selenium was first suggested to be essential in 1957, this was not firmly established until a biochemical role was identified for selenium in 1972. The first report of human selenium deficiency appeared in 1979; the subject resided in a low-selenium area and was receiving total parenteral nutrition (TPN) after surgery. The subject and other selenium-deficient subjects on TPN exhibited bilateral muscular discomfort, muscle pain, wasting, and cardiomyopathy. Subsequently, it was discovered that Keshan disease, prevalent in certain parts of China, was prevented by selenium supplementation. Keshan disease is a multiple focal myocardial necrosis resulting in acute or chronic heart function insufficiency, heart enlargement, arrhythmia, pulmonary edema, and death. Other consequences of inadequate selenium include impaired immune function and increased susceptibility to viral infections. Selenium deficiency also can make some non-virulent viruses become virulent.

Recently, however, not only selenium deficiency, but effects of supranutritional intakes of selenium have become of great health interest. Several supplementation trials have indicated that selenium has anticarcinogenic properties. For example, one trial with 1312 patients supplemented with either 200>g selenium/day or with a placebo found the selenium treatment was statistically associated with reductions in several types of cancer including colorectal and prostate cancers.

Selenium is a relatively toxic element; intakes averaging 1.2 mg/day can induce changes in nail structure. Chronic selenium intakes over 3.2 mg/day can result in the loss of hair and nails, mottling of the teeth, lesions in the skin and nervous system, nausea, weakness, and diarrhea.

Selenium, which is biologically important as an anion, is homeostatically regulated by excretion, primarily in the urine but some also is excreted in the breath. Selenate, selenite, and selenomethionine are all highly absorbed by the GI tract; absorption percentages for these forms of selenium are commonly found to be in the 80 to 90 percent range.

Food sources of selenium are fish, eggs, and meat from animals fed abundant amounts of selenium and grains grown on high-selenium soil.

Boron

Recent findings with this trace element suggest that it may be of nutritional importance, although a clearly defined biochemical function for boron in higher animals and humans has not been defined. It has been hypothesized, however, that boron has a role in cell membrane function that influences the response to hormones, transmembrane signaling, or transmembrane movement of regulatory cations or anions. Human studies suggest that a low boron intake can impair cognitive and psychomotor function and the inflammatory response, as well as increasing the susceptibility to osteoporosis and arthritis.

About 85 percent of ingested boron is absorbed and excreted in the urine shortly after ingestion. Because boron homeostasis is regulated efficiently by urinary excretion, it is a relatively non-toxic element. A tolerable upper level intake of 20 mg/day was determined for boron by the Food and Nutrition Board of the United States National Academy of Sciences.

An analysis of both human and animal data by a World Health Group suggested that an acceptable safe range of population mean intakes for boron for adults could be 1 to 13 mg/day. Foods of plant origin, especially fruits, leafy vegetables, nuts, pulses, and legumes are rich sources of boron.

Chromium

A naturally occurring biologically active form of chromium called chromodulin has been described that apparently has a role in carbohydrate and lipid metabolism as part of a novel insulin-amplification mechanism. Chromodulin is an oligopeptide that binds four chromic ions and facilitates insulin action in converting glucose into lipids and carbon dioxide.

The nutritional importance of chromium is currently a controversial subject. Chromium deficiency has been suggested to impair glucose tolerance, which could eventually lead to diabetes. Supranutritional chromium supplementation (1,000 >g/day) has been found beneficial for some cases of type II diabetes. Supplements containing supranutritional amounts of chromium in the picolinate form have been promoted as being able to induce weight loss and to increase muscle mass. However, most ergogenic (work output) oriented studies have found chromium picolinate supplementation to be ineffective for increasing muscle mass, strength, and athletic performance, and there are no data from well-designed studies to support the claim that chromium picolinate supplementation is an effective weight loss modality. Chromium in the +3 valency state is a relatively non-toxic element.

Chromium homeostasis is regulated by intestinal absorption, which is low. Estimates of absorption range from less than 0.5 to 2 percent. Absorbed chromium is excreted in the urine.

The Food and Nutrition Board of the United States Academy of Sciences determined that there was not sufficient evidence to set an estimated average requirement of chromium. Therefore, an adequate intake was set based on estimated mean intakes. The adequate intake for young males was set at 35>g/day, and that for young females was set at 25>g/day. Some of the best food sources of chromium are whole grains, pulses, some vegetables (for example, broccoli and mushrooms), liver, processed meats, ready-to-eat cereals, and spices.

It is likely that not all the essential mineral elements for humans have been identified. Moreover, numerous biochemical functions for trace elements most likely remain to be identified. Thus, the full extent of the pathological consequences of marginal or deficient intakes of the trace elements has not been established. Furthermore, some trace elements such as selenium, fluoride, and lithium in supranutritional amounts are being found to have therapeutic or preventative value against disease. Thus, the determination of the importance of trace elements for human health and well-being should be considered a work in progress with some exciting advances likely in the future.

Nutritionally Required Trace Elements

New Trace Elements

Boron, chromium, manganese, nickel, tin, vanadium, molybdenum, arsenic, lithium, aluminium, strontium, cesium and silicon are regarded as new trace elements in the sense that they have only recently been considered essential in human diets. These elements are the subject of exciting research in animals, particularly ruminants, where they have been shown to be essential in one or more species. For example, ruminants feeding on grass grown in soil where molybdenum levels are abnormally high have demonstrated an increased tendency to exhibit copper deficiency. However, for many of these new trace elements, such as manganese, there is no evidence at the present time that abnormally low or high dietary intakes cause substantial nutritional problems in human populations.

Non nutritional Trace Elements as Environmental Contaminants

The application of different non-invasively collected matrices for the detection and quantification of environmental exposure to metals has abundantly been described. Iyengar and Rapp illustrate how the placenta can be used to detect the presence of toxic trace elements such as arsenic, mercury or lead. Others additionally provide data indicating that concentrations of trace elements can vary 5- to 10-fold due to specific exposure situations, such as living in the vicinity of coal and metal mining and smelting operations. Also urine, finger/toenails, or human milk have repeatedly been used as suitable matrices of exposure for metals. Probably the best-known use of hair as a non-invasive matrix for metals is in the biomonitoring of organic and inorganic mercury, as hair is by far the best integrator of past exposure, although also other metals have repeatedly been monitored using hair.

Also for organic compounds, several non-invasively collected matrices have frequently been used to quantify chemicals that are lipophylic and resistant to metabolic degradation. Many persistent environmental pollutants, such as polychlorinated dibenzo-p-dioxins (PCDDs) and dibenzofurans (PCDFs) are widespread and can be found in placenta, cord blood, or human milk. Meconium has been promoted as a very useful non-invasively collected matrix to describe exposure of the unborn child to pesticides or metals.

Urine has frequently been used as a matrix for a wide variety of both organic and inorganic compounds for occupational as well as environmental exposure. As urine is a main excretory pathway, it is a preferred matrix to monitor for non-bioaccumulating and rapidly metabolised compounds. Interpretation of urinary biomarkers is sometimes complicated as they primarily focus on quantification of metabolites rather than the parent compounds. Some chemicals, like chlorpyrifos, are broken down in the body into a number of metabolites which can be detected in urine. As these same metabolites also occur as natural products of environmental degradation, it is not always possible to distinguish exposure to the parent compound from exposure to its environmental degradates.

The gaseous fraction of exhaled breath can be a good biomonitor for a wide variety of volatile substances. The monitoring of metals in exhaled breath is less well-documented, although Mutti et. al showed that toxic metals and transition elements are detectable in exhaled breath condensate (EBC). This may have potential in the assessment of the target tissue dose for substances with potential pneumotoxic activity, such as Cd, Co or Ni. A disadvantage of measuring biomarkers of exposure in the gaseous phase of exhaled air is that the residence time of substances are generally rather short, in the order of minutes or hours.

When using hair and finger/toenails to quantify exposure to environmental contaminants, caution is needed in the interpretation of exposure data. Many contaminants have been proven to reach hair and finger/toenails via two major routes: endogenous (xenobiotics reach the hair matrix through blood) and exogenous (atmospheric deposition). Hence it may be difficult to distinguish between contaminants taken up and those related to external contamination. Findings show that hair can in fact be used for the quantification of exogenous atmospheric exposure and, in some cases, even for the estimation of corresponding air concentrations. Mainly for substances where environmental exposure does not generally occur through air, the constraint of external contamination is limited. For methylmercury, for example, which is generally taken up through food or dental amalgam, hair is by far the preferred matrix for biomonitoring and large-scale biomonitoring campaigns which have clearly linked methylmercury with neurodevelopmental deficiencies. While washing procedures have been shown to differentiate between endogenous and exogenous exposure, caution remains necessary in the interpretation of using hair as a non-invasively collected matrix for exposure.

Biomarkers of Effect

Detecting the early hazardous effects of environmental contaminants has a high priority in the protection of neonates and newborn children. In both placental tissue and cord blood, biomarkers for DNA damage have been measured using different methods, including P-postlabeling and enzyme-linked immunosorbent assay (ELISA) methods. Although the estimated polycyclic aromatic hydrocarbon (PAH) dose to the foetus may be 10-fold lower than in the mothers because of the earlier mentioned barrier effect of the placenta, the PAH adduct levels in the newborns are similar or higher than those in their mothers. This may imply that the foetus is far more susceptible to DNA damage than its mother. Maervoet et. al outlined that, while there are still many gaps in the understanding of the relationship between environmental contaminants in cord blood and the functioning of the thyroid system, any interference may adversely affect neonatal neurodevelopment. The discovery that placental nucleic acids can be used as a marker for prenatal screening has recently opened the door for the application of rapidly developing technologies such as toxicogenomics and proteomics. This in the future may lead to the development of molecular markers for non-invasive prenatal gene expression profiling of the foetus using placenta and cord blood.

Also for other non-invasively collected matrices, the use of relatively new -'omics' technologies has opened the door for the detection of new biomarkers of effect. Recently, there has been increased research interest in the metabolic profiling (metabonomics or metabolomics) of mainly urine and exhaled breath samples.

This technique encompasses the systematic profiling of metabolite concentrations and their systematic and temporal changes through effects from diet, lifestyle, environment and genetic. Although metabonomics currently is not yet sufficiently developed for large-scale biomonitoring studies, it may be a promising tool for the rapid screening of metabolites.

Other non-invasively collected matrices already have a long-standing and well-documented history as a matrix in routine, often clinical, health assessment. For example, α_1- and β_2-microglobulin excretion and retinol-binding proteins in urine have frequently been described as sensitive biomarkers of renal disfunctioning. Additionally, urine has been used as a non-invasive matrix for the quantification of base DNA-adducts as biomarkers for carcinogenesis. Urinary 8-hydroxy-deoxyguanosine (8-OHdG) has been used as a biomarker of the DNA repair response to oxidative stress and DNA-damaging compounds. Exhaled breath has received much attention as a suitable clinical matrix for the early detection of pulmonary and respiratory diseases. It has extensively been demonstrated that fractional exhaled nitric oxide (FENO) concentrations in exhaled air are higher in people suffering from various pulmonary diseases, including asthma and chronic obstructive pulmonary disease (COPD). Other biomarkers, such as exhaled breath temperature, pH of EBC, and the presence of cytokines (e.g. interleukins (IL) IL-4, IL-6, IL-8 and tumor-necrosis factor-α (TNF-α)), 8-isoprostane or hydrogen peroxide may also be biomarkers of lung inflammation and oxidative stress. Saliva as a non-invasively collected matrix has been used to quantify cholinesterase activity, a biomarker for potential neurotoxic effects. Additionally, saliva specimens have been used in combination with the Ames test or a Chinese hamster V79 lung fibroblast cell line to investigate the genotoxic effects of smoking and alcohol consumption. Exfoliated buccal cells have been used to monitor genetic damage in humans using the micronucleus test.

For other non-invasively collected matrices, such as hair, human milk, finger/toenails or meconium, we found no examples of biomarkers of effect.

Correlations among (Non-invasive Collected) Biomonitoring Matrices

Many publications have measured and compared values for biomarkers of exposure for the same chemicals in different (non-invasively collected) matrices. Often, there are good correlations among invasively collected and non-invasively collected matrices, although several exceptions can be noticed.

In several cases, meconium has been shown to be the most sensitive matrix to analyse fetal exposure to environmental contaminants. For example, in mercury or pesticides monitoring, meconium showed the highest exposure rates compared to cord blood and infant hair. For phthalate metabolites, urinary concentrations have been found to be more informative than blood, serum or milk concentrations in the Swedish population. Urinary concentrations showed lower day-to-day variability and were detected at much higher concentrations than in other matrices.

For other contaminants, good correlations are often found between paired cord blood or placenta samples and other matrices such as maternal blood, human milk or amniotic fluid.

Urinary biomarkers generally reflect metabolites of compounds rather than the pure compound itself. This makes it more difficult to correlate urinary biomarkers with biomarkers of exposure in other matrices. Mainly for non-metabolised compounds, correlations between concentrations in urine and other matrices such as blood appear to be good. Many authors also have addressed the correlation between contaminants in hair and other matrices. For inorganic compounds, Mehra and Juneja reported significant correlations between hair and fingernail concentrations of Cd and Pb, Stupar et. al found significant correlations between Pb in hair and blood, and Ng et. al reported significant correlations between hair mercury and mercury in blood, 24 h urine or cord blood samples following a meta-analysis of mercury biomonitoring studies.

Compared to invasively collected matrices such as peripheral blood, there are very little, if any, chemicals that cannot be determined with sufficient sensitivity in non-invasively collected matrices. Concentration of chemicals in non-invasively collected matrices generally correlates very well with other, often invasively collected, matrices such as peripheral blood.

Validation Status

Validation of biomarkers is a lengthy and difficult process, that requires the fulfilment of multiple criteria. In general, these criteria include understanding the biological and temporal relevance, pharmacokinetics, background variability, exposure-dose or dose-response relationship and identification of confounding factors.

Several urinary biomarkers have been fully validated for organic and inorganic substances, reference materials and internationally recognised standard methodologies are often available, and sampling procedures are well-documented. It is a widely employed non-invasively collected matrix in many of the largest environmental studies such as the German Environmental Survey for children (GerES), the American National Health and Nutrition Examination Study (NHANES), or the Flemish Human Biomonitoring programme. Also human milk as a matrix has widespread use for biomarkers of exposure. It is easily collectable and fully validated for a wide variety of substances and reference materials. Internationally recognised standards for analysis are often available, and sampling procedures are well-documented. The WHO co-ordinates a regular exposure study to using human milk to quantify the global presence and distribution of persistent organic pollutants. Within the context of this repeated human biomonitoring program, guidelines for developing a national protocol for sampling human milk have been developed. Methods of collecting and sampling exhaled breath biomarkers include Tedlar bags, canisters, or even portable real-time detectors.

These methods are generally well validated and documented, as are the measurement techniques.. Because of the frequent application of meconium for illicit drug screening, structured screening protocols have also been developed for this non-invasively collected matrix.

Slotnick et. al evaluated the effect of demographic characteristics and nutritional measures on the association between drinking water and toenail arsenic concentrations. Apart from toenail iron concentration, no demographic or nutritional parameters affected the biomarker response. In Kenya, however, Were et. al reported that socio-economic background, health conditions, dietary habits and urban or rural living had a significant effect on lead, cadmium, zinc and iron concentrations in the fingernails of school children. For arsenic, Brima and co-workers illustrated that different ethnic groups showed significantly different concentrations of total arsenic in fingernails, suggestive of a different pattern of arsenic metabolism in different ethnic groups.

Chapter 10

Heavy Metal Toxicity

1. Metabolism, absorption, distribution, excretion and mechanism of toxicity for each of the metals.
2. Characteristics of a good biological chelating agent.
3. Uses of chelating agents in terms of:
   a. Route of administration
   b. Mechanism of action
   c. Specificity
   d. Stability of the heavy metal-chelator complex
   e. Toxicity of chelator and the complex.

METALS AND DRUGS (CHELATORS)

METAL CHELATING AGENTS (DRUGS)

Lead- Ethylenediamine-tetraacetic acid (EDTA)

2,3-dimercaptosuccinic acid (Succimer)

2,3-dimercaptopropanol (BAL, Dimercaprol)

Penicillamine

Cadmium- Ethylenediamine-tetraacetic acid (EDTA)

Mercury- N-acetyl-penicillamine (NAP)

Penicillamine

2,3-dimercaptopropanol (BAL, Dimercaprol)

2,3-dimercaptosuccinic acid (Succimer)

Arsenic- N-acetyl-penicillamine (NAP)

Antimony- Ethylenediamine-tetraacetic acid (EDTA)

Iron - Deferoxamine

HEAVY METALS AND ANTIDOTES

I. METALS

A. Lead (Pb)

1. Absorption

(i) Skin: alkyl lead compounds, because of lipid solubility (methyl and tetraethyl lead)

(ii) Inhalation: up to 90% depending upon particle size

(iii) GI: adults 5 to 10%, children 40%

2. Distribution

Initially carried in red cells and distributed to soft tissues (kidney and liver); redistributed to bone, teeth and hair mostly as a phosphate salt. Rates of absorption and distribution are greatly influenced by dietary intake and body stores of phosphate, calcium and iron relative to lead

(i) high PO_4, Pb storage in bone
(ii) high Vitamin D, Pb storage in soft tissue
(iii) low PO_4, Pb sequestered in soft tissue
(iv) high Ca++, Pb sequestered in soft tissue

Half life in blood: 30-60 days, bone: 20-30 years

3. Sources of exposure

(i) GI - paint, pottery, moonshine
(ii) Inhalation - metal fumes
(iii) Skin-tetraethyl lead in gasoline

4. Mechanisms of toxicity

(i) Inhibition of heme biosynthesis (see figure 1). Heme is the essential structural component of hemoglobin, myoglobin and cytochromes

(ii) Binds to sulfhydryl groups (-SH groups) of proteins

5. Diagnosis

(1) History of exposure

(2) Whole blood lead level
- Children: >25 µg/dl treatments
- Adults: >50 µg/dl candidates for treatment; > 80 µg/dl & symptomatic, treatment initiated, >120 µg/dl encephalopathy

(3) Protoporphyrin levels in erythrocytes are usually elevated with lead levels > 40 µg/dl

(4) Urinary lead excretion >80 ìg/dl

(5) Urinary delta aminolevulonic acid (ALA)

(6) Lead mobilization test (redistribution toxicity)

6. Symptoms

(1) Acute - nausea, vomiting, thirst, diarrhea/constipation, abdominal pain, hemoglobinuria, oliguria leading to hypovolemic shock

(2) Chronic -

+ GI: lead colic (nausea, vomiting, abdominal pain)
+ NMJ: lead palsy (fatigue, wrist-drop)
+ CNS: lead encephalopathy (headache, vertigo, irritation, insomnia, CNS edema)

7. Treatment

(1) Remove from exposure

(2) CaNa2EDTA

(3) 2,3-dimercaptopropanol (Dimercaprol, BAL)

(4) 2,3-dimercaptosuccinic acid (Succimer)

(5) D-penicillamine

MECHANISM OF LEAD TOXICOLOGY:
HEME BIOSYNTHESIS

Succinyl CoA + Glycine

Pb ⟶ delta-Aminolevulinate synthetase

↓

delta Aminoevulinate

Pb ⟶ delta-Aminolevulinate dehydratase

↓

Porphobilinogen

Porphobilinogen deaminase
Urophorphyrinogen III cosynthase

↓

Uroporphyrinogen III

Urophorphyrinogen decarboxylase

↓

Coproporphyrinogen III

Pb ⟶ Coproporphyrinogen oxidase

↓

Protoporphyrin IX

Pb ⟶ Ferrochelatase + Fe (+2)

↓

HEME

Figure 1.

B. Cadmium (Cd++)

1. Absorption

(1) Inhalation 10 to 40%

(2) GI 1.5 to 5%

2. Distribution:

Initially bound to albumin and blood cells; subsequently bound to metallothionnin in liver and kidney tissue. Half life is 10-20 years

3. Sources of exposure

(1) GI - pigments, polishes, antique toys

(2) Environmental - electroplating, galvanization, plastics, batteries

(3) Inhalation industrial, metal fumes, tobacco - 1-2 μg/pack

4. Mechanisms of toxicity

(1) Inhalation: lung - local irritation and inhibition of alpha1-antitrypsin associated with emphysema

(2) Oral: kidney-proximal tubular injury (proteinuria) associated with beta2-macroglobulin

5. Diagnosis

(1) History of exposure

(2) Whole blood cadmium level >80 ìg/dl

6. Symptoms

a. Acute

+ Oral: vomiting, diarrhea, abdominal cramps

+ Inhalation: chest pains, nausea, dizziness, diarrhea, pulmonary edema

b. Chronic

+ Oral: nephrotoxicity

+ Inhalation: emphysema-like syndrome and Nephrotoxicity

7. Treatment

(1) Remove from exposure

(2) CaNa2EDTA

(3) 2,3-dimercaptopropanol (BAL), Cadmium complex is extremely nephrotoxic and therefore is not used

C. Mercury (Hg)

1. Absorption

(1) GI: inorganic salts are variably absorbed (10%) but may be converted to organic mercury (methyl and ethyl in the gut by bacteria), organic compounds are well absorbed >90%

(2) Inhalation: elemental Hg completely absorbed

2. Distribution depends upon sources of exposure

Elemental Hg (vapor) crosses membranes well and rapidly moves from the lung to the CNS. Organic salts (lipid-soluble) are evenly distributed, intestinal (intracellular) fecal elimination. Inorganic salts concentrate in blood, plasma and kidney (renal elimination).

Half life is 60 to 70 days.

3. Sources of exposure

(1) Environmental from electronics and plastic industry

(2) Seed fungicide treatment, dentistry

4. Mechanisms of toxicity

(1) Dissociation of salts precipitates proteins and destroys mucosal membranes

(2) Necrosis of proximal tubular epithelium

(3) Inhibition of sulfhydryl (-SH) group containing enzymes

5. Diagnosis

(1) History of exposure

(2) Blood mercury

6. Symptoms

(1) Acute

(a) (inorganic salts) degradation of mucosa- GI pain, vomiting, diuresis, anemia, hypovolemic shock, renal toxicity.

(b) (organic) CNS involvement- vision, depression, irritability, blushing, intention tremors, insomnia, fatigue, diuresis.

(2) Chronic: CNS symptoms similar to acute organic poisoning with gingivitis, tachycardia, goiter, increased urinary Hg.

7. Treatment

(1) Remove from exposure

(2) Hg and Hg salts > 4 µg/dl : 2,3-dimercaptopropanol (BAL), $,$-dimethyl cysteine (penicillamine), most effective is N-acetyl-$,$-dimethyl cysteine (N-acetyl-penicillamine)

(3) Methyl Hg- supportive treatment (non-absorbable thiol resins can be given orally to reduce methyl Hg level in the gut).

D. Arsenic, As^{3+}, As^{5+}

1. Absorption

(1) GI inorganic: trivalent and pentavalent salts >90%. organic: also bound as tri and pentavalent >90%

(2) Inhalation: uptake is dependent upon particle size

2. Distribution

Accumulates in lung, heart, kidney, liver, muscle and neural tissue.

Concentrates in skin, nails and hair. Half life is 7 to 10 hours.

3. Sources of exposure

(1) GI - well water, food. Environmental: by-product of smelting ore, AsGa in semiconductors, herbicides and pesticides

(2) Inhalation - fumes and dust from smelting

4. Mechanisms of toxicity

(1) Membranes: protein damage of capillary endothelium increased vascular permeability leading to vasodilation and vascular collapse

(2) Inhibition of sulfhydryl group containing enzymes

(3) Inhibition of anaerobic and oxidative phosphorylation (substitutes for inorganic phosphate in synthesis of high-energy phosphates)

5. Diagnosis

(1) History of exposure

(2) Blood and urinary levels (acute)

(3) Hair or fingernail (chronic)

6. Symptoms

(1) Acute - damage to mucosa, sloughing, hypovolemic shock, fever, GI discomport/pain, anorexia

(2) Chronic - weakness, GI, hepatomegaly (jaundice > cirrhosis), melanosis, arrhythmias, peripheral neuropathy, peripheral vascular disease (blackfoot disease)

(3) Carcinogenicity - epidemiologic evidence; liver angiosarcoma, skin and lung cancer

7. Treatment

(1) Remove from exposure

(2) Acute: supportive therapy: fluid, electrolyte replacement, blood pressure support (dopamine)

(3) Chronic: penicillamine without dialysis

+ Arsine gas (AsH_3) acts as a hemolytic agent with secondary to renal failure.

+ supportive therapy: transfusion; **(chelators have not been shown to be beneficial)**

E. Antimony

1. Absorption

skin, inhalation and Gastrointestinal

Ceramics, batteries, toys, drugs (organic antimonials)

2. Mechanisms

Inhibition of SH-group containing enzymes and substrate phosphorylation

3. Diagnosis and Symptoms - same as for Arsenic

4. Treatment

(1) Remove from exposure

(2) BAL and penicillamine, EDTA

(3) Stibine gas (SbH3) is a hemolytic agent-supportive treatment

II. CHELATING AGENTS

Chelation is the formation of a metal ion complex in which the metal ion is associated with a charged or uncharged electron donor referred to as a ligand. It usually attaches or co-ordinates using one, two or more electron pair donor atoms (oxygen, nitrogen and sulfur).

A. Requirements for the ideal chelating agent

1. the compound should be soluble in aqueous medium
2. the chelating agent should be stable in the circulation
3. if it is given orally, it should be absorbed by the GI tract and it should be cleared by the kidney
4. the compound should be active at physiological pH
5. the compounds should chelate only the specific metals
6. the chelator itself should not be toxic
7. the chelator-metal complex should be less toxic than the metal alone

B. Chelators

1. 2,3-dimercaptopropanol (dimercaprol) also known as British Anti Lewisite (BAL)- given IMin peanut oil

a. **Structure**

Isolated

$$\begin{matrix} CH_2 & - & CH & - & CH_2 \\ | & & | & & | \\ SH & & SH & & OH \end{matrix}$$

Complexed

$$\begin{matrix} CH_2 & - & CH & - & CH_2 \\ | & & | & & | \\ S & & S & & OH \\ & \diagdown & & \diagup & \\ & & Hg & & \end{matrix}$$

$$\begin{matrix} CH_2 & - & CH & - & CH_2 \\ | & & | & & | \\ S & & S & & OH \\ & \diagdown & \diagup & & \\ & & Hg & & \\ & \diagup & \diagdown & & \\ S & & S & & OH \\ | & & | & & \diagdown \\ CH_2 & - & CH & - & CH_2 \end{matrix}$$

b. **Use**: arsenic, mercury, antimony, lead, gold, zinc, bismuth. Half life is less than one hour.

c. **Toxicity**:

i. CNS convulsions in high dose, given Q4H (i.e., given every 4 hrs)

ii. Increased Blood pressure (50 mm Hg) due to tachycardia and peripheral constriction of arterioles.

iii. Renal toxicity can be reduced by alkalizing the urine, which protects against dissociation of the metal-BAL complex. Toxicity is 50%

+ Anxiety
+ increased blood pressure
+ nausea, vomiting and headache
+ 1/3 of children treated react with fever

2. Ethylene diamine-tetraacetic acid (EDTA) given IV as the Calcium disodium salt.

Uses

i. Disodium EDTA binds calcium in blood to prevent clotting- used in blood collection and storage

ii. chelator for lead

iii. only chelates circulating metal because EDTA cannot enter inside the cell membrane

iv. frequently used in combination with BAL or penicillamine for treatment of lead poisoning

c. **Toxicity:**

i. tubular destruction due to release of metal or the metal-EDTA complex in the kidney.

ii. thrombophlebitis in some cases– too rapid infusion

3. d-isomer of beta,beta-dimethylcysteine (Penicillamine) given orally

a. Structure

Isolated | Complexed

$$H_3C-\underset{SH}{\overset{CH_3}{C}}-\underset{NH_2}{CH}-COOH$$

$$H_3C-\underset{S\to Hg}{\overset{CH_3}{C}}-\underset{H\to Hg}{CH}-COOH$$

N-acetyl penicillamine

$$H_3C-\underset{SH}{\overset{CH_3}{C}}-CH(-N-C(=O)-CH_3)-C(=O)-OH$$

$$H_3C-\overset{CH_3}{C}(-S\to Hg)-CH(-N\to Hg)-C(=O\to Hg)-OH;\quad N-C(=O)-CH_3$$

b. Uses:

i. lead, mercury, arsenic

ii. copper - Wilson's disease (reduced or monovalent copper is elevated)

c. Toxicity:

i. l-isomer depletes pyridoxine (Vitamin B6), therefore d-isomer is used clinically

ii. fever, skin rashes, leukopenia, nausea, vomiting

iii. anaphylactic reactions (do not use in the case allergic to penicillin)

4. Deferoxamine

a. Structure

Isolated | Complexed

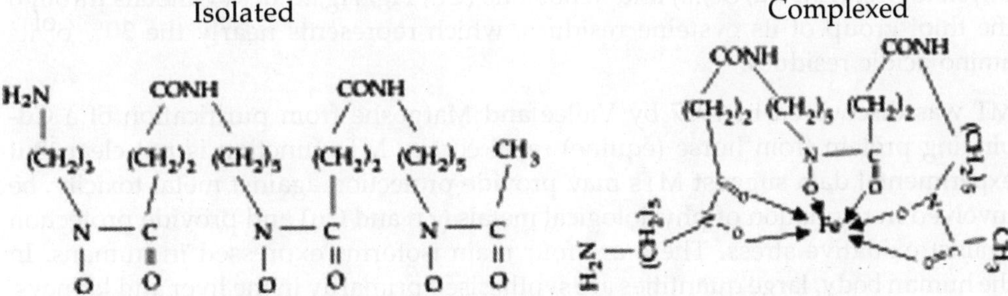

b. Uses: iron (Fe^{3+}) poisoning IM or slow IV. It has also been used orally to chelate iron poisoning

c. **Toxicity:**

i. skin rash

ii. histamine release with reduced blood pressure (shock)

iii. cataracts

5. **Trientine (triethylenetetramine HCl)** is a polydentate chelating agent. It chelates copper and less toxic than penicillamine, but it is teratogenic in long-term use. It is used for treating wilson's disease (hepatolenticular degeneration).

a. **Structure**

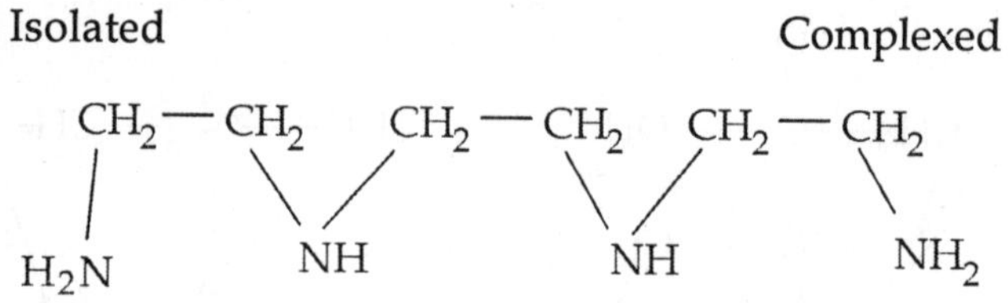

Isolated Complexed

$$\mathrm{H_2N{-}CH_2{-}CH_2{-}NH{-}CH_2{-}CH_2{-}NH{-}CH_2{-}CH_2{-}NH_2}$$

b. Uses:

i. lead, mercury, arsenic

ii. copper - Wilson's disease (reduced or monovalent copper is elevated)

c. Toxicity:

i. l-isomer depletes pyridoxine (Vitamin B6), therefore d-isomer is used clinically

ii. fever, skin rashes, leukopenia, nausea, vomiting

iii. anaphylactic reactions (do not use in the case of allergic to penicillin)

Metallothionein

Metallothionein (MT) is a family of cysteine-rich, low molecular weight (MW ranging from 3500 to 14000 Da) proteins. MTs have the capacity to bind both physiological (Zn, Cu, Se,...) and xenobiotic (Cd, Hg, Ag,...) heavy metals through the thiol group of its cysteine residues, which represents nearly the 30% of its amino acidic residues.

MT was discovered in 1957 by Vallee and Margoshe from purification of a Cd-binding protein from horse (equine) renal cortex. MTs function is not clear, but experimental data suggest MTs may provide protection against metal toxicity, be involved in regulation of physiological metals (Zn and Cu) and provide protection against oxidative stress. There are four main isoforms expressed in humans. In the human body, large quantities are synthesised primarily in the liver and kidneys.

Their production is dependent on availability of the dietary minerals, as zinc, copper and selenium, and the amino acids histidine and cysteine.

Structure and Classification

MTs are present in a vast range of taxonomic groups, ranging from prokaryotes (such as the cyanobacteria *Syneccococus spp*....), protozoa (p. ex. the ciliate *Tetrahymena* genera...), plants (such as *Pisum sativum, Triticum durum, Zea mays, Quercus suber*...), yeast (such as *Saccharomyces cerevisiae, Candida albicans*,...), invertebrates (such as the nematode *Caenorhabditis elegans*, the insect *Drosophila melanogaster*, the mollusc *Mytilus edulis*, or the echinoderm *Strongylocentrotus purpuratus*) and vertebrates (such as the chicken, *Gallus gallus*, or the mammalian *Homo sapiens* or *Mus musculus*).

The MTs from this diverse taxonomic range represent a high-heterogeneity sequence (regarding molecular weight and number and distribution of Cysteine residues) and do not show general homology; in spite of this, homology is found inside some taxonomic groups (such as vertebrate MTs).

From their primary structure, MTs have been classified by different methods. The first one dates from 1987, when Fowler *et al*., established three classes of MTs: Class I, including the MTs which show homology with horse MT, Class II, including the rest of the MTs with no homology with horse MT, and Class III, which includes phytochelatins, Cysteine-rich enzymatically synthesised peptides which are no longer considered MTs. The second classification was performed by Binz and Kagi in 2001, and takes into account taxonomic parameters and the patterns of distribution of Cysteine residues along the MT sequence. It results in a classification of 15 families. Family 15 contains the plant MTs, which in 2002 have been further classified by Cobbet and Goldsbrough into 4 Types (1, 2, 3 and 4) depending on the distribution of their Cysteine residues and a Cysteine-devoid regions (called spacers) characteristic of plant MTs.

A Table including the principal aspects of the two latter classifications is included.

Family	Sequence pattern	Example
1.Vertebrate	K-x(1,2)-C-C-x-C-C-P-x(2)-C	*M.musculus* MT1 MDPNCSCTTGGSCACAGSCKCKECKCTSCKKCCSCCPV GCAKCAQGCVCKGSSEKCRCCA
2.Molluscan	C-x-C-x(3)-C-T-G-x(3)-C-x-C-x(3)-C-x-C-K	*M.edulis* 10MTIV MPAPCNCIETNVCICDTGCSGEGCRCGDACKCSGADCK CSGCKVVCKCSGSCACEGGCTGPSTCKCAPGCSCK
3.Crustacean	P-[GD]-P-C-C-x(3,4)-C-x-C	*H.americanus* MTH MPGPCCKDKCECAEGGCKTGCKCTSCRCAPCEKCTSGC KCPSKDECAKTCSKPCKCCP
4.Echinoderms	P-D-x-K-C-[V,F)-C-C-x(5)-C-x-C-x(4)-C-C-x(4)-C-C-x(4,6)-C-C	*S.purpuratus* SpMTA MPDVKCVCCKEGKECACFGQDCCKTGECCKDGTCCGI CTNAACKCANGCKCGSGCSCTEGNCAC

5.Diptera	C-G-x(2)-C-x-C-x(2)-Q-x(5)-C-x-C-x(2)D-C-x-C	*D.melanogaster* MTNB MVCKGCGTNCQCSAQKCGDNCACNKDCQCVCKNGP KDQCCSNK
6.Nematoda	K-C-C-x(3)-C-C	*C.elegans* MT1 MACKCDCKNKQCKCGDKCECSGDKCCEKYCCEEASEK KCCPAGCKGDCKCANCHCAEQKQCGDKTHQHQGTA AAH
7.Ciliate	x-C-C-C-x	*T.termophila* MTT1 MDKVNSCCCGVNAKPCCTDPNSGCCCVSKTDNCCKSD TKECCTGTGEGCKCVNCKCCKPQANCCCGVNAKPCCF DPNSGCCCVSKTNNCCKSD TKECCTGTGEGCKCTSCQCCKPVQQGCCCGDKAKACC TDPNSGCCCSNKANKCCDATSKQECQTCQCCK
8.Fungal 1	C-G-C-S-x(4)-C-x-C-x(3,4)-C-x-C-S-x-C	*N.crassa* MT MGDCGCSGASSCNCGSGCSCSNCGSK
9.Fungal 2	---	*C.glabrata* MT2 MANDCKCPNGCSCPNCANGGCQCGDKCECKKQSCHG CGEQCKCGSHGSSCHGSCGCGDKCECK
10.Fungal 3	---	*C.glabrata* MT2 MPEQVNCQYDCHCSNCACENTCNCCAKPACACTNSAS NECSCQTCKCQTCKC
11.Fungal 4	C-X-K-C-x-C-x(2)-C-K-C	*Y.lipolitica* MT3 MEFTTAMLGASLISTTSTQSKHNLVNNCCCSSSTSESSMP ASCACTKCGCKTCKC
12.Fungal 5	---	*S.cerevisiae* CUP1 MFSELINFQNEGHECQCQCGSCKNNEQCQKSCSCPTGC NSDDKCPCGNKSEETKKSCCSGK
13.Fungal 6	---	*S.cerevisiae* CRS5 TVKICDCEGECCKDSCHCGSTCLPSCSGGEKCKCDHSTG SPQCKSCGEKCKCETTCTCEKSKCNCEKC
14.Procaryota	K-C-A-C-x(2)-C-L-C	Synechococcus sp SmtA MTTVTQMKCACPHCLCIVSLNDAIMVDGKPYCSEVCA NGTCKENSGCGHAGCGCGSA
15.Plant		
15.1.Plant MTs Type 1	C-X-C-X(3)- C-X-C-X(3)- C-X-C-X(3) spacer C-X-C-X(3)- C-X-C-X(3)- C-X-C-X(3)	*Pisum sativum* MT MSGCGCGSSCNCGDSCKCNKRSSGLSYSEMETTETVILG VGPAKIQFEGAEMSAASEDGGCKCGDNCTCDPCNCK
15.2.Plant MTs Type 2	C-C-X(3)-C-X-C-X(3)- C-X-C-X(3)- C-X-C-X(3)-spacer- C-X-C-X(3)- C-X-C-X(3)- C-X-C-X(3)	*L.esculetum* MT MSCCGGNCGCGSSCKCGNGCGGCKMYPDMSYTESSTT TETLVLGVGPEKTSFGAMEMGESPVAENGCKCGSDCKC NPCTCSK
15.3.Plant MTs Type 3	---	*A.thaliana* MT3 MSSNCGSCDCADKTQCVKKGTSYTFDIVETQESYKEAMI MDVGAEENNANCKCKCGSSCSCVNCTCCPN

15.4.Plant MTs Type 4 or Ec	C-x(4)-C-X-C-X(3)-C-X(5)-C-X-C-X(9,11)-HTTCGCGEHC-X-C-X(20)-CSCGAXCNCASC-X(3,5)	*T.aestium* MT MGCNDKCGCAVPCPGGTGCRCTSARSDAAAGEHTTCG CGEHCGCNPCACGREGTPSGRANRRANCSCGAACNCA SCGSTTA

Secondary structure elements have been observed in several MTs SmtA from *Syneccochoccus*, mammalian MT3, Echinoderma SpMTA, fish *Notothenia Coriiceps* MT, Crustacean MTH, , but until this moment, the content of such structures is considered to be poor in MTs, and its functional influence is not considered.

The **tertiary level of structure** show also a high hetereogeneity. While vertebrate, echinoderm and crustacean MTs show a bidominial structure with divalent metals as Zn(II) or Cd(II) (the protein is folded binding metals in two functionally independent domains, with a metallic *cluster* each), yeast and procariotyc MTs show a monodominial structure (only one domain as a receipt of a unique metallic *cluster*). In spite of no structural data is available for molluscan, nematoda and Drosophila MTs, it is commonly assumed that the formers are bidominial and the latter mondominial. No conclusive data are disposable for Plant MTs, but two possible tridimensional structures have been proposed: (1) a bidominial structure similar to that of vertebrate MTs; (2) a codominial structure, in which two Cys-rich domains interacts to form a unique metallic cluster.

Quaternary level of structure has been scarcely taken in account for MTs. Dimerization and oligomerization processes has been observed and attributed to several molecular mechanisms: (1) the formation of inter-molecular disulfide bridges by the oxidation of thiol residues. (2) the formation of a metallic bridge by a metal with bounds to Cys belonging to different MTs. (3) the formation of a metallic bridge by a metal with bounds to His belonging to different MTs. (4) the interaction through inorganic phosphate ions. In spite of dimeric and polymeric MTs have shown to acquire novel properties on metal detoxification, its physiological significance has been demonstrated only in the case of procaryotic Synechococcus SmtA, which forms an MT dimer able to form structures similar to Zn-fingers which develops Zn-regulatory activities.

Metallothioneins present also diverse metal-binding preferences, which have been associated to functional specificity. As an example, the mammalian *Mus musculus* MT1 presents a preference for binding divalent metal ions (Zn(II), Cd(II),...), while yeast CUP1 presents a preference for binding monovalent metal ions (Cu(I), Ag(I),...). Taking as a reference from these functional preferences a **novel functional classification** of MTs which classifies MTs as Zn- or Cu-thioneins is currently being developed.

Function

Metal-binding

Metallothionein has been documented to bind a wide range of metals including cadmium, zinc, mercury, copper, arsenic, silver, etc. Metallation of MT was previously been reported to occur co-operatively but recent reports have provided strong evidence that metal-binding occur via a sequencial non-cooperative mechanism. The observation of partially-metallated MT (as in not fully saturated MT) suggest that these species are biologically important.

Metallothionein likely participates in the uptake, transport, and regulation of zinc in biological systems. Mammalian MT binds three Zn(II) in the beta domain and four Zn(II) in the alpha domain. Cysteine is a sulfur-containing amino acid, hence the name -"thionein" (thio means sulfur). However, the participation of inorganic sulfide and chloride ions has been proposed for some MT forms. In some MTs, mostly bacterial, histidine participates in zinc binding. By binding and releasing zinc, metallothioneins (MTs) may regulate zinc level within the body. Zinc, in turn, is a key element for the activation and binding of certain transcription factors through its participation in the (aptly-named) zinc finger region of the protein. Metallothionein also carries zinc ions (signals) from one part of the cell to another. When zinc enters a cell, it can be picked up by thionein (which thus becomes "metallothionein") and carried to another part of the cell where it is released to another organelle or protein. In this way the thionein-metallothionein becomes a key component of the zinc signaling system in cells. This system is particularly important in the brain, where zinc signaling is prominent both between and within nerve cells. It also seems to be important for the regulation of the tumor suppressor protein p53.

Control of Oxidative Stress

Cysteine residues from MTs can capture harmful oxidant radicals like the superoxide and hydroxyl radicals. In this reaction, cysteine is oxidized to cystine, and the metal ions which were bound to cysteine are liberated to the media. As explained in the *Expression and regulation* section, this Zn can activate the synthesis of more MTs. This mechanism has been proposed to be an important mechanism in the control of the oxidative stress by MTs. The role of MTs in oxidative stress has been confirmed by MT Knockout mutants, but some experiments propose also a prooxidant role for MTs.

Expression and Regulation

Metallothionein gene expression is induced by a high variety of stimuli, as metal exposure, oxidative stress, glucocorticoids, hydric stress, etc. The level of the response to these inducers depends on the MT gene. MT genes present in their promotors specific sequences for the regulation of the expression, elements as Metal Response Elements (MRE), Glucocorticoid Response Elements (GRE),...

Metallothionein and Disease

Cancer

Because MTs play an important role in transcription factor regulation, problems with MT function or expression may lead to malignant transformation of cells and ultimately cancer. Studies have found increased expression of MTs in some cancers of the breast, colon, kidney, liver, lung, nasopharynx, ovary, prostate, mouth, salivary gland, testes, thyroid and urinary bladder; they have also found lower levels of MT expression in hepatocellular carcinoma and liver adenocarcinoma.

There is evidence to suggest that higher levels of MT expression may also lead to resistance to chemotherapeutic drugs.

Autism

It has been hypothesized that a MT disorder explains several symptoms of autism (the mercury poisoning hypothesis), but a 2006 study found that autistic children did not differ significantly from normal children in levels of MT or antibodies to MT.

Heat Shock Proteins

Heat shock proteins are an evolutionary conserved family of proteins whose expression increases in response to a variety of different metabolic results. Despite their designation, most of the heat shock proteins are constitutively expressed and perform essential functions. Most notable is their role as molecular chaperones, facilitating the synthesis and folding of proteins throughout the cell. In addition, heat shock proteins have been shown to participate in protein assembly, secretion, trafficking, protein degradation, and the regulation of transcription factors and protein kinases. Increased levels of heat shock proteins after stress plays a central role in cellular homeostasis.

Heat shock proteins (HSPs), also called stress proteins, are a group of proteins that are present in all cells in all life forms. They are induced when a cell undergoes various types of environmental stresses like heat, cold and oxygen deprivation.

Heat shock proteins are also present in cells under perfectly normal conditions. They act like 'chaperones', making sure that the cell's proteins are in the right shape and in the right place at the right time. For example, HSPs help new or distorted proteins fold into shape, which is essential for their function. They also shuttle proteins from one compartment to another inside the cell, and transport old proteins to 'garbage disposals' inside the cell. Heat shock proteins are also believed to play a role in the presentation of pieces of proteins (or peptides) on the cell surface to help the immune system recognize diseased cells. For decades it has been known that animals can be 'vaccinated' against cancer. This is how it works: Tumor cells can be weakened, or attenuated, and injected like a vaccine into a mouse. Afterwards, if these same tumor cells, at full strength, are injected into the mouse, the mouse will reject the tumor cells and cancer will not develop. However, if a mouse has not been vaccinated in this manner, the tumor cells will 'take' and the mouse will develop cancer.

Although, it was clear that animals could be vaccinated against cancer for a long time, it was not known how it worked. Then about 25 years ago, a graduate student named Pramod Srivastava began a series of experiments. He took tumor cells, broke them open, and separated the different parts of the cells into fractions. He then used each of the fractions as 'vaccines' to see which fraction protected the mice from developing cancer. After many experiments, he found that the element responsible for protecting the mice was heat shock proteins.

Heat shock proteins trigger immune response through activities that occur both inside the cell (*intracellular*) and outside the cell (*extracellular*).

Intracellular Activities

Because of the normal functions of heat shock proteins inside the cell (such as helping proteins fold, preparing proteins for disposal, etc.), HSPs end up binding virtually every protein made within the cell. This means that at any given time, HSPs can be found inside the cell bound to a wide array of peptides that represent a 'library' of all the proteins inside the cell. This library contains normal peptides that are found in all cells as well as abnormal peptides that are only found in sick cells.

Research suggests that inside the cell, heat shock proteins take the peptides and hand them over to another group of molecules. These other molecules take the abnormal peptides that are found only in sick cells and move them from inside the cell to outside on the cell's surface. When the abnormal peptides are displayed in this way, they act as red flags, warning the immune system that the cell has become sick. These abnormal peptides are called *antigens* – a term that describes any substance capable of triggering an immune response.

Extracellular Activities

Heat shock proteins are normally found inside cells. When they are found outside the cell, it indicates that a cell has become so sick that it has died and spilled out all of its contents. This kind of messy, unplanned death is called necrosis, and only occurs when something is very wrong with the cell. Extracellular HSPs are one of the most powerful ways of sending a 'danger signal' to the immune system in order to generate a response that can help to get rid of an infection or disease.

How does Antigenics' Heat Shock Protein Technology Work?

Antigenics' heat shock protein technology works by mimicking the 'danger signal' believed to be naturally triggered by extracellular HSPs. Depending on the abnormal peptides contained within the HSP-associated 'library' of proteins that have spilled out of the cell, the immune system can be activated to target different cancers and certain infectious agents.

Antigenics' investigational patient-specific cancer vaccine Oncophage® (vitespen; formerly HSPPC-96) consists of HSP-peptide complexes that have been isolated from individual patient's cancer cells. Because cancer is so incredibly variable, the abnormal peptides found within diseased cells are different from cancer to cancer and from person to person. Therefore, this library of abnormal peptides is unique to each individual's disease and can be thought of as the cancer's 'fingerprint'.

When the vaccine is injected into the body, the fingerprint of HSP-peptide complexes can directly encounter the immune system's cells, which is designed to stimulate the immune cells to target cancer cells bearing this fingerprint.

Chapter 11

Cytotoxicity, Genotoxicity And Carcinogenesis

Cytotoxicity

Cytotoxicity is the quality of being toxic to cells. Examples of toxic agents are a chemical substance, an immune cell or some types of venom (e.g. from the puff adder or brown recluse spider).

Cell Physiology

Treating cells with a cytotoxic compound can result in a variety of cell fates. The cells may undergo necrosis, in which they lose membrane integrity and die rapidly as a result of cell lysis. The cells can stop actively growing and dividing (a decrease in cell viability), or the cells can activate a genetic program of controlled cell death (apoptosis).

Cells undergoing necrosis typically exhibit rapid swelling, lose membrane integrity, shut down metabolism and release their contents into the environment. Cells that undergo rapid necrosis *in vitro* do not have sufficient time or energy to activate apoptotic machinery and will not express apoptotic markers. Apoptosis is characterized by well-defined cytological and molecular events including a change in the refractive index of the cell, cytoplasmic shrinkage, nuclear condensation and cleavage of DNA into regularly-sized fragments. Cells in culture that are undergoing apoptosis eventually undergo secondary necrosis. They will shut down metabolism, lose membrane integrity and lyse.

Measuring cytotoxicity

Cytotoxicity assays are widely used by the pharmaceutical industry to screen for cytotoxicity in compound libraries. Researchers can either look for cytotoxic compounds, if they are interested in developing a therapeutic that targets rapidly dividing cancer cells, for instance; or they can screen "hits" from initial high-throughput drug screens for unwanted cytotoxic effects before investing in their development as a pharmaceutical.

Assessing cell membrane integrity is one of the most common ways to measure cell viability and cytotoxic effects. Compounds that have cytotoxic effects often compromise cell membrane integrity. Vital dyes, such as trypan blue or propidium iodide are normally excluded from the inside of healthy cells; however, if the cell membrane has been compromised, they freely cross the membrane and stain intracellular components. Alternatively, membrane integrity can be assessed by monitoring the passage of substances that are normally sequestered inside cells to the outside. One commonly measured molecule is lactate dehydrogenase (LDH). Protease biomarkers have been identified that allow researchers to measure relative numbers of live and dead cells within the same cell population. The live-cell protease is only active in cells that have a healthy cell membrane, and loses activity once the cell is compromised and the protease is exposed to the external environment. The dead-cell protease cannot cross the cell membrane, and can only be measured in culture media after cells have lost their membrane integrity.

Cytotoxicity can also be monitored using the MTT or MTS assay. This assay measures the reducing potential of the cell using a colorimetric reaction. Viable cells will reduce the MTS reagent to a colored formazan product. A similar redox-based assay has also been developed using the fluorescent dye, resazurin. In addition to using dyes to indicate the redox potential of cells in order to monitor their viability, researchers have developed assays that use ATP content as a marker of viability. Such ATP-based assays include bioluminescent assays in which ATP is the limiting reagent for the luciferase reaction.

Cytotoxicity can also be measured by the sulforhodamine B (SRB) assay, WST assay and clonogenic assay.

A label-free approach to follow the cytotoxic response of adherent animal cells in real-time is based on electric impedance measurements when the cells are grown on gold-film electrodes. This technology is referred to as electric cell-substrate impedance sensing (ECIS). Label-free real-time techniques provide the kinetics of the cytotoxic response rather than just a snapshot like in many colorimetric endpoint assays.

Immune System Cytotoxicity

Antibody-dependent cell-mediated cytotoxicity (*ADCC*) describes the cell-killing ability of certain lymphocytes, which requires the target cell being marked by an antibody. *Lymphocyte-mediated cytotoxicity*, on the other hand, does not have to be mediated by antibodies; nor does *complement-dependent cytotoxicity* (*CDC*), which is mediated by the complement system.

Three groups of cytotoxic lymphocytes are distinguished:

- Cytotoxic T cells
- Natural killer cells
- Natural killer T cells

Genotoxicity

Genotoxicity describes a deleterious action on a cell's genetic material affecting its integrity. Genotoxic substances are known to be potentially mutagenic or carcinogenic, specifically those capable of causing genetic mutation and of contributing to the development of tumors. This includes both certain chemical compounds and certain types of radiation.

Typical genotoxins like aromatic amines are believed to cause mutations because they are nucleophilic and form strong covalent bonds with DNA resulting with the formation of Aromatic Amine-DNA Adducts, preventing accurate replication.

Genotoxins affecting sperm and eggs can pass genetic changes down to descendants who have never been exposed to the genotoxin.

The fish along with other aquatic organisms play an important role in maintaining the aquatic ecosystem and also constitute important source of food for the human beings. The aquatic environments remains the ultimate receipt of increasing number of anthropogenic chemicals, the large portion of which are potentially genotoxic and mutagenic, which can interact and damage DNA even at sublethal/ non-lethal concentrations.

Genotoxicity refers to potentially harmful effects of pollutants on genetic material indicated by induced damage to DNA without direct evidence of mutation.

Mutagenicity refers to the induction of permanent transmissible changes that may involve a single gene or gene segments, a block of genes, or whole chromosomes.

The genotoxic/mutagenic chemicals responsible for DNA damage in fishes may cause malignancies, reduced fertility, malformations, reduced survival of embryos and larvae (tetrogenesis), reduced viability and growth , thereby not only reducing species biodiversity but also affect the economy of fish production significantly. Thus, there is growing concern worldwide, over the presence of a variety of anthropogenic contaminants that cannot only adversely affect the aquatic organisms but also pose risk to human health via food chain. With a view to assess the genotoxicity caused by these pollutants, several biomarkers have been developed for fishes that can be used for environmental biomonitoring.

The genotoxic chemicals effect at different organizational levels of fish

DNA	DNA strand breakages, mutations
Cell	Micronuclei formation, necrosis/apoptosis, disordered proliferation and differentiation
Tissue/Organ	Altered gene/protein expression, altered metabolism, neoplastic transformation
Individual	Hereditary defects and malformations, tumors
Populations	Reduced viability, fertility and population size
Aquatic ecosystem	Reduced species diversity

The need of genotoxicity studies in fishes:

- Fish constitutes an important and cheap source of animal proteins to human beings and a large number of people depend on fish/fishing activities for their livelihood.

- Due to intensive development of industries, agricultural and urban activities, especially along the seashores and riverbanks, thousands of new compounds and organic chemicals are being deposited into the nature water bodies.
- As these pollutants contain potential genotoxic/mutagenic chemicals, there is a growing concern regarding the occurrence of malignancies in aquatic organisms and hereditary defects through mutations in germ cells.
- The mutagenic/genotoxic compounds might induce substantial reproductive loss in exposed populations and could further influence individual fitness.
- From ecotoxicology point of view, fish serves as an important biological indicator of water quality and can highlight the potential dangers of new chemicals introduced in the aquatic environment.
- Fish play different roles in the trophic web such as undergoing bioaccumulation of environmental pollutants, biotransformation of xenobiotics through cytochrome P450 dependent oxidative metabolism.
- Fish can respond to mutagens at low concentration of toxicants in a manner similar to higher vertebrates and DNA repair slower in fishes than mammals.
- Owing to the above reasons, fish is also being increasingly used as a test model for eco toxicological studies.

Biomarkers for assessment of genotoxic potential in fishes:

1. Single cell gel electrophoresis (SCGE) or comet assay : Detects single/double breaks in the sugar-phosphate backbone of the DNA in the individual.
2. Sister chromatid exchanges (SCEs): Detects reciprocal exchanges of DNA between two sister chromatids of a duplicating chromosome. The exchange process presumably involves DNA breakage and repair.
3. Unscheduled DNA repair synthesis after excision and removal of a stretch of DNA containing a region of damage induced by chemical substances or physical agents.

Biomarkers for detection of mutagenic effects in germ cells and somatic cells

1. Chromosome aberration test (CAT) : Detects change in chromosome structure as observed by microscope examination of the metaphase stage of cell division (structural abberations) and also change in the number of chromosomes from the normal number characteristic of the cells.
2. Micronucleus assay (MNT) : Detects induction of micronucleus formation i.e., small nucleus, separate from and additional to the main nucleus, produced during telophase of mitosis (meiosis) by lagging chromosome fragments or whole chromosome.

Other important non-conventional biomarkers:

1. Induction of the enzymatic Ethoxy Resorufin-O-Deethylase (EROD) following exposure to Polycyclic Aromatic Hydrocarbons (PAHs).
2. Random amplification polymorphic DNA (RAPD) can be used to detect DNA alterations such as DNA adducts formation, breaks, and point mutations, large rearrangements induced by chemical or physical agents following direct and/or indirect interaction with DNA.
3. The combination of DNA and fluorescent *in situ* hybridization (FISH) techniques is useful to explore the part of gene that is being affected.
4. DNA microarray technology has increased the possibilities of analyzing the chemicals and environmental stressors on large number genes in a simple experiment.

Future prospects of genetic toxicology:

1. The measurement of gene expression levels upon exposure to a chemical would help to evaluate the impact of anthropogenic stress and the potential of the fishes to adapt and survive in changing environments.
2. DNA microarray technology has ample possibilities of analyzing the chemicals and environmental stressors on large number genes in a simple experiment.
3. These modern technologies along with conventional biomarkers can play a pivotol role in monitoring the effect of pollutants on aquatic organisms and to help framing proper environmental policies to safeguard our rich aquatic biodiversity and ecosystem health.

Carcinogenicity

The term *carcinogen* refers to any substance, radionuclide or radiation that is an agent directly involved in the promotion of cancer or in the increase of its propagation. This may be due to the ability to damage the genome or to the disruption of cellular metabolic processes. Several radioactive substances are considered carcinogens, but their carcinogenic activity is attributed to the radiation, for example, gamma rays and alpha particles, which they emit. Common examples of carcinogens are inhaled asbestos, certain dioxins, and tobacco smoke.

Cancer is a disease where damaged cells of the patient's body do not undergo programmed cell death, but their growth is no longer controlled and their metabolism is altered. Carcinogens may increase the risk of getting cancer by altering cellular metabolism or damaging DNA directly in cells, which interferes with biological processes, and induces the uncontrolled, malignant division, ultimately leading to the formation of tumors. Usually DNA damage, if too severe to repair, leads to programmed cell death, but if the programmed cell death pathway is damaged, then the cell cannot prevent itself from becoming a cancer cell.

There are many natural carcinogens. Aflatoxin B_1, which is produced by the fungus *Aspergillus flavus* growing on stored grains, nuts and peanut butter, is an example of a potent, naturally-occurring microbial carcinogen. Certain viruses such as Hepatitis B and human papilloma viruses have been found to cause cancer in humans. The first one shown to cause cancer in animals is Rous sarcoma virus, discovered in 1910 by Peyton Rous.

Benzene, kepone, EDB, asbestos, and the waste rock of oil shale mining have all been classified as carcinogenic. As far back as the 1930s, industrial smoke and tobacco smoke were identified as sources of dozens of carcinogens, including benzo[*a*]pyrene, tobacco-specific nitrosamines such as nitrosonornicotine, and reactive aldehydes such as formaldehyde—which is also a hazard in embalming and making plastics. Vinyl chloride, from which PVC is manufactured, is a carcinogen and thus a hazard in PVC production.

Co-carcinogens are chemicals that do not necessarily cause cancer on their own, but promote the activity of other carcinogens in causing cancer.

After the carcinogen enters the body, the body makes an attempt to eliminate it through a process called biotransformation. The purpose of these reactions is to make the carcinogen more water-soluble so that it can be removed from the body. But these reactions can also convert a less toxic carcinogen into a more toxic one.

DNA is nucleophilic, therefore soluble carbon electrophiles are carcinogenic, because DNA attacks them. For example, some alkenes are toxicated by human enzymes to produce an electrophilic epoxide. DNA attacks the epoxide, and is bound permanently to it. This is the mechanism behind the carcinogenity of benzo[*a*]pyrene in tobacco smoke, other aromatics, aflatoxin and mustard gas.

CERCLA identifies all radionuclides as carcinogens, although the nature of the emitted radiation (alpha, beta, gamma, or neutron and the radioactive strength), its consequent capacity to cause ionization in tissues, and the magnitude of radiation exposure, determine the potential hazard. Carcinogenicity of radiation depends on the type of radiation, type of exposure and penetration. For example, alpha radiation has low penetration and is not a hazard outside the body, but are carcinogenic when inhaled or ingested.

For example, Thorotrast, a (incidentally-radioactive) suspension previously used as a contrast medium in x-ray diagnostics, is a potent human carcinogen known because of its retention within various organs and persistent emission of alpha particles. Marie Curie, one of the pioneers of radioactivity, died of cancer caused by radiation exposure during her experiments.

Not all types of electromagnetic radiation are, in fact, carcinogenic. Low-energy waves on the electromagnetic spectrum are generally not including radio waves, microwave radiation, infrared radiation and visible light. Higher-energy radiation, including ultraviolet radiation (present in sunlight), x-rays, and gamma radiation, generally *is* carcinogenic, if received in sufficient doses.

Several published studies suggest a link between exposure to light at night and risk of breast cancer, due to suppression of the normal nocturnal production of melatonin. In 1978, Cohen et al. proposed that reduced production of the hormone melatonin might increase the risk of breast cancer and citing "environmental lighting" as a possible causal factor. Researchers at the National Cancer Institute (NCI) and National Institute of Environmental Health Sciences have concluded a study that suggests that artificial light during the night can be a factor for breast cancer. A good review of current knowledge of the health consequences of exposure to artificial light at night and an explanation of the causal mechanisms has been published in the Journal of Pineal Research in 2007.

Substances or foods irradiated with electrons or electromagnetic radiation (such as microwave, X-ray or gamma) are not carcinogenic. No "radiation" remains, just like no light remains in room after you turn out the light. (In contrast, non-electromagnetic neutron radiation produced inside nuclear reactors can produce secondary radiation by making bombarded substances radioactive.)

In other words, there is no possibility of getting cancer if the food is consumed.

Carcinogens in Prepared Food

Cooking food at high temperatures, for example, grilling or barbecuing meats, can lead to the formation of minute quantities of many potent carcinogens that are comparable to those found in cigarette smoke (i.e., benzo[*a*]pyrene). Charring of food resembles coking and tobacco pyrolysis, and produces similar carcinogens. There are several carcinogenic pyrolysis products, such as polynuclear aromatic hydrocarbons, which are converted by human enzymes into epoxides, which attach permanently to DNA. Pre-cooking meats in a microwave oven for 2–3 minutes before grilling shortens the time on the hot pan, and removes heterocyclic amine (HCA) precursors, which can help minimize the formation of these carcinogens.

Reports from the Food Standards Agency have found that the known animal carcinogen acrylamide is generated in fried or overheated carbohydrate foods (such as french fries and potato chips). Studies are underway at the FDA and European regulatory agencies to assess its potential risk to humans. The charred residue on barbecued meats has been identified as a carcinogen, along with many other tars.

Casein, a milk protein, may also be a carcinogen as it is linked to the promotion of cancer and other diseases which was discovered in the 1980s by nutrition and health researcher, Dr. T. Colin Campbell, author of The China Study. Casein is often listed as sodium caseinate, calcium caseinate or milk protein. These are often found in energy bars, drinks as well as packaged goods.

Nevertheless, the fact that the food contains minute quantities does not necessarily mean that there is a significant hazard, although eliminating exposure may decrease ones odds of getting cancer, the gastrointestinal tract sheds its outer layer continuously to protect itself from carcinomas, and has a high activity of detoxifying enzymes.

Circadian Disruption

"Shiftwork that involves circadian disruption" was listed, in 2007, as a probable carcinogen by the World Health Organization's International Agency for Research on Cancer. Multiple studies have documented a link between night shift work and the increased incidence of breast cancer. Circadian disruption by exposure to light at night suppresses the production of the hormone melatonin which leads to reduction in cellular immune defense and surveillance necessary for protection from development of cancers. Melatonin also seems to have a direct protective effect against cancer possibly in part because of its strong antioxidant properties.

Mechanisms of Carcinogenicity

Carcinogens can be classified as genotoxic or non-genotoxic. Genotoxins cause irreversible genetic damage or mutations by binding to DNA. Genotoxins include chemical agents like N-nitroso-N-methylurea (MNU) or non-chemical agents such as ultraviolet light and ionizing radiation. Certain viruses can also act as carcinogens by interacting with DNA.

Non-genotoxins do not directly affect DNA but act in other ways to promote growth. These include hormones and some organic compounds.

Procarcinogen

A procarcinogen is a precursor to a carcinogen. One example is nitrites when taken in by the diet. They are not carcinogenic themselves, but turn into nitrosamines in the body, which are carcinogenic.

Common Carcinogens

Occupational Carcinogens

Occupational carcinogens are agents that pose a risk of cancer in several specific work-locations:

Carcinogen	Associated cancer sites or types	Occupational uses or sources
Arsenic and its compounds	• Lung • Skin • Hemangiosarcoma	• Smelting by product • Component of: • Alloys • Electrical and semiconductor devices • Medications (e.g. melarsoprol) • Herbicides • Fungicides • Animal dips

Asbestos	• Lungs • Mesothelioma • Gastrointestinal tract	Not in use, but still found in: • Constructions • Roofing papers • Floor tiles • Fire-resistant textiles • Friction linings
Benzene	• Leukemia • Hodgkin lymphoma	• Light fuel oil • Former use as solvent and fumigant • Printing • Lithography • Paint • Rubber • Dry cleaning • Adhesives • Coatings • Detergents
Beryllium and its compounds	• Lung	• Missile fuel • Light-weight alloys • Aerospace applications • Nuclear reactors
Cadmium and its compounds	• Prostate	• Yellow pigments • Phosphors • Solders • Batteries • Metal paintings and coatings
Hexavalent chromium(VI) compounds	• Lung	• Paints • Pigments • Preservatives
Ethylene oxide	• Leukemia	• Ripening agent for fruits and nuts • Rocket propellant • Fumigant for foodstuffs and textiles • Sterilant for hospital equipment
Nickel	• Nose • Lung	• Nickel plating • Ferrous alloys • Ceramics • Batteries • Stainless-steel welding by-product
Radon and its decay products	• Lung	• Uranium decay • Quarries and mines • Cellars and poorly ventilated places

Vinyl chloride	• Hemangiosarcoma • Liver	• Refrigerant • Vinyl polymers • Adhesive for plastics • Former use in pressurized containers
Shiftwork that involves circadian disruption	• Breast	–
Involuntary smoking (Passive smoking)	• Lung	• Tobacco smoke

Others

- Gasoline (contains aromatics)
- Lead and its compounds
- Alkylating antineoplastic agents (chemotherapy)
- Ultraviolet radiation from the sun and UV lamps

MUTAGENS AND CARCINOGENESIS

In biology, a mutagen (Latin, literally *origin of change*) is a physical or chemical agent that changes the genetic material (usually DNA) of an organism and thus increases the frequency of mutations above the natural background level. As many mutations cause cancer, mutagens are typically also carcinogens. Not all mutations are caused by mutagens: so-called "spontaneous mutations" occur due to errors in DNA replication, repair and recombination.

Effects of Mutation

The changes in nucleic acid sequences by mutations include substitution of nucleotide base-pairs and insertions and deletions of one or more nucleotides in DNA sequences. Although some of these mutations are lethal, or cause serious disease, many have minor effects, as the changes they cause in the sequence of encoded proteins are not significant. Many mutations cause no visible effects at all, either because they occur in introns or because they do not change the amino-acid sequence, due to redundancy of codons.

Genetic Drift

The change in a population's genetic material due to the accumulation of random chance is called drift, and serves as a molecular clock. In general, the more nucleotide differences between two organisms, the more time has elapsed since their last common ancestor. Though it is difficult to determine in many organisms, estimates for mutation rates have been made for both *E. coli* and eukaryotes. It was estimated that, in these organisms, about one nucleotide in every 10^{10} is changed, and continues through reproduction to future generations of cells.

History/Discovery of Mutagenesis

In the 1920s, Hermann Muller discovered that x-rays caused mutations in fruit flies. He went on to use x-rays to create Drosophila mutants that he used in his studies of genetics. He also discovered that x-rays not only mutate genes in fruit flies but also have effects on the genetic makeup of humans. The first mutagens to be identified were carcinogens, or cancer-causing substances. Early physicians detected tumors in patients more than 2,000 years before the discovery of chromosomes and DNA. In 500 B.C., the Greek Hippocrates named crab-shaped tumors cancer (meaning crab).

In England in 1775, Dr. Percivall Pott wrote a paper on the high incidence of scrotal cancer in chimney sweeps who were typically boys small enough to fit inside chimneys and clean out the soot. Pott suggested that chimney soot contained carcinogens that could cause the growth of the warts seen in scrotal cancer. Over 150 years later, chimney soot was found to contain hydrocarbons capable of mutating DNA.

In France in the 1890s, Bordeaux wine workers showed an unusually high incidence of skin cancer on the back of the neck. These workers spend their days bending over in the fields picking grapes, exposing the back of their necks to the sun. The ultraviolet (UV) radiation in natural sunlight was later identified as a mutagen.

Nature of Mutagens

Mutagens are usually chemical compounds or ionizing radiation. Mutagens can be divided into different categories according to their effect on DNA replication:

- Some mutagens act as base analogs and get inserted into the DNA strand during replication in place of the substrates.
- Some react with DNA and cause structural changes that lead to miscopying of the template strand when the DNA is replicated.
- Some work indirectly by causing the cells to synthesize chemicals that have the direct mutagenic effect.

The Ames test is one method to determine how mutagenic an agent is.

Examples

- Ionizing radiation, for example, X-rays, gamma rays and alpha particles
- UV light
- Base analogs, which can substitute for DNA bases and cause copying errors
- Deaminating agents such as Nitrous acid

- Intercalating agents such as Ethidium bromide
- Alkylating agents such as Ethylnitrosourea
- Transposons, sections of DNA that undergo autonomous fragment relocation/multiplication
- Some natural plant alkaloids, such as those from Vinca species.
- Bromine and some of its compounds
- Sodium azide
- Psoralen combined with ultraviolet radiation causes DNA crosslinking and hence chromosome breakage
- Benzene

Mutagens in Fiction

In science fiction, mutagens are often represented as substances that are capable of completely changing the form of the recipient.

Carcinogenesis

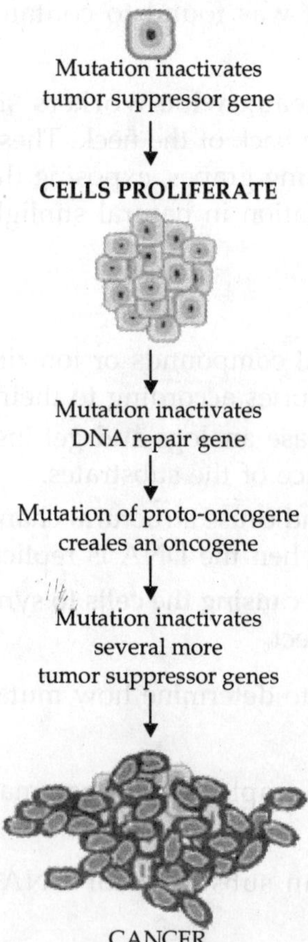

Cancers are caused by a series of mutations. Each mutation alters the behavior of the cell somewhat.

Carcinogenesis (*the creation of cancer)*, is the process by which normal cells are transformed into cancer cells.

Cell division is a physiological process that occurs in almost all tissues and under many circumstances. Under normal circumstances, the balance between proliferation and programmed cell death, usually in the form of apoptosis, is maintained by tightly regulating processes to ensure the integrity of organs and tissues. Mutations in DNA that lead to cancer (only certain mutations can lead to cancer and the majority of potential mutations will have no bearing) disrupt these orderly processes by disrupting the programming regulating the processes.

Carcinogenesis is caused by this mutation of the genetic material of normal cells, which upsets the normal balance between proliferation and cell death. This results in uncontrolled cell division and the evolution of those cells by natural selection in the body. The uncontrolled and often rapid proliferation of cells can lead to benign tumors; some types of these may turn into malignant tumors (cancer). Benign tumors do not spread to other parts of the body or invade other tissues, and they are rarely a threat to life unless they compress vital structures or are physiologically active; for instance, producing a hormone. Malignant tumors can invade other organs, spread to distant locations (metastasis) and become life-threatening.

More than one mutation is necessary for carcinogenesis. In fact, a series of several mutations to certain classes of genes is usually required before a normal cell will transform into a cancer cell. Only mutations in those certain types of genes which play vital roles in cell division, apoptosis (cell death), and DNA repair will cause a cell to lose control of its cell proliferation.

Mechanisms of Carcinogenesis

Cancer is a genetic disease. In order for cells to start dividing uncontrollably, genes that regulate cell growth must be damaged. Proto-oncogenes are genes that promote cell growth and mitosis, whereas tumor suppressor genes discourage cell growth, or temporarily halt cell division to carry out DNA repair. Typically, a series of several mutations to these genes is required before a normal cell transforms into a cancer cell.

Proto-oncogenes

Proto-oncogenes promote cell growth in a variety of ways. Many can produce hormones, "chemical messengers" between cells that encourage mitosis, the effect of which depends on the signal transduction of the receiving tissue or cells. Some are responsible for the signal transduction system and signal receptors in cells and tissues themselves, thus controlling the sensitivity to such hormones. They often produce mitogens, or are involved in transcription of DNA in protein synthesis, which create the proteins and enzymes responsible for producing the products and biochemicals, cells use and interact with them.

Mutations in proto-oncogenes can modify their expression and function, increasing the amount or activity of the product protein. When this happens, they become oncogenes, and thus, cells have a higher chance to divide excessively and uncontrollably. The chance of cancer cannot be reduced by removing proto-oncogenes from the genome, as they are critical for growth, repair and homeostasis of the body. It is only when they become mutated that the signals for growth become excessive. It is important to note that a gene possessing a growth-promoting role may increase carcinogenic potential of a cell, under the condition that all necessary cellular mechanisms that permit growth are activated. This condition includes also the inactivation of specific tumor suppressor genes. If the condition is not fulfilled, the cell may cease to grow and can proceed to die. This makes knowledge of the stage and type of cancer cell that grows under the control of a given oncogene crucial for the development of treatment strategies.

Tumor Suppressor Genes

Tumor suppressor genes code for anti-proliferation signals and proteins that suppress mitosis and cell growth. In general, tumor suppressors are transcription factors that are activated by cellular stress or DNA damage. Often DNA damage will cause the presence of free-floating genetic material as well as other signs, and will trigger enzymes and pathways that lead to the activation of tumor suppressor genes. The functions of such genes is to arrest the progression of cell cycle in order to carry out DNA repair, preventing mutations from passing on to daughter cells. Canonical tumor suppressors include the p53 gene, which is a transcription factor activated by many cellular stresses including hypoxia and ultraviolet radiation damage.

However, a mutation can damage the tumor suppressor gene itself, or the signal pathway that activates it, switching it off. This can hinder DNA repair, so that DNA damage accumulates, including changes that lead to cancer.

Multiple Mutations

In general, mutations in both types of genes are required for cancer to occur. For example, a mutation limited to one oncogene would be suppressed by normal mitosis control and tumor suppressor genes, first hypothesised by the Knudson hypothesis. A mutation to only one tumor suppressor gene would not cause cancer either, due to the presence of many "backup" genes that duplicate its functions. It is only when enough proto-oncogenes have mutated into oncogenes, and enough tumor suppressor genes are deactivated or damaged, that the signals for cell growth overwhelm the signals to regulate it, that cell growth quickly spirals out of control. Often, because these genes regulate the processes that prevent most damage to genes themselves, the rate of mutations increases as one gets older, because DNA damage forms a feedback loop.

Usually, oncogenes are dominant alleles, as they contain gain-of-function mutations, whereas mutated tumor suppressors are recessive alleles, as they contain loss-of-function mutations. Each cell has two copies of a same gene, one from each parent, and, under most cases, gain-of-function mutation in one copy of a particular proto-oncogene is enough to make that gene a true oncogene, while usually loss-of-function mutation must happen in both copies of a tumor suppressor gene to render that gene completely non-functional. However, cases exist in which one loss of function copy of a tumor suppressor gene can render the other copy non-functional, called the dominant negative effect. This is observed in many p53 mutations.

Mutation of tumor suppressor genes that are passed on to the next generation of not merely cells, but their offspring, can cause increased likelihoods for cancers to be inherited. Members within these families have increased incidence and decreased latency of multiple tumors. The mode of inheritance of mutant tumor suppressors is that affected member which inherits a defective copy from one parent, and a normal copy from another. Because mutations in tumor suppressors act in a recessive manner (however, there are exceptions), the loss of the normal copy creates the cancer phenotype. For instance, individuals that are heterozygous for p53 mutations are often victims of Li-Fraumeni syndrome, and that are heterozygous for Rb mutations develop retinoblastoma. In similar fashion, mutations in the adenomatous polyposis coli gene are linked to adenopolyposis colon cancer, with thousands of polyps in the colon while young, whereas mutations in BRCA1 and BRCA2 lead to early onset of breast cancer.

Non-mutagenic Carcinogens

Many mutagens are also carcinogens, but some carcinogens are not mutagens. Examples of carcinogens that are not mutagens include alcohol and estrogen. These are thought to promote cancers through their stimulating effect on the rate of cell mitosis. Faster rates of mitosis increasingly leave less opportunities for repair enzymes to repair damaged DNA during DNA replication, increasing the likelihood of a genetic mistake. A mistake made during mitosis can lead to the daughter cells' receiving the wrong number of chromosomes, which leads to aneuploidy and may lead to cancer.

Role of Viral Infections

Furthermore, many cancers originate from a viral infection; this is especially true in animals such as birds, but less so in humans. 12% of human cancers can be attributed to a viral infection. The mode of virally-induced tumors can be divided into two, acutely-transforming or slowly-transforming. In acutely-transforming viruses, the viral particles carry a gene that encodes for an overactive oncogene called viral-oncogene (v-onc), and the infected cell is transformed as soon as v-onc is expressed. In contrast, in slowly-transforming viruses, the virus genome is inserted, especially as viral genome insertion is obligatory part of retroviruses, near a proto-oncogene in the host genome. The viral promoter or other transcription regulation elements, in turn, cause over-expression of that proto-oncogene, which in turn, induces uncontrolled cellular proliferation. Because viral genome insertion is not specific to proto-oncogenes and the chance of insertion near that proto-oncogene is low, slowly-transforming viruses have very long tumor latency compared to acutely-transforming virus, which already carries the viral-oncogene.

How Viruses are Thought to Cause Cancer

Viruses that are known to cause cancer such as HPV (cervical cancer), Hepatitis B (liver cancer), and EBV (a type of lymphoma), are all DNA viruses. It is thought that when the virus infects a cell, it inserts a part of its own DNA near the cell growth genes, causing cell division. The group of changed cells that are formed from the first cell dividing all have the same viral DNA near the cell growth genes. The group of changed cells are now special because one of the normal controls on growth has been lost.

Depending on their location, cells can be damaged through radiation from sunshine, chemicals from cigarette smoke, and inflammation from bacterial infection or other viruses. Each cell has a chance of damage, a step on a path toward cancer. Cells often die if they are damaged, through failure of a vital process or the immune system; however, sometimes damage will knock out a single cancer gene. In an old person, there are thousands, tens of thousands or hundreds of thousands of knocked-out cells. The chance that any one would form a cancer is very low.

When the damage occurs in any area of changed cells, something different occurs. Each of the cells has the potential for growth. The changed cells will divide quicker when the area is damaged by physical, chemical, or viral agents. A vicious circle has been set up: Damaging the area will cause the changed cells to divide, causing a greater likelihood that they will suffer knock-outs.

This model of carcinogenesis is popular because it explains why cancers grow. It would be expected that cells that are damaged through radiation would die or at least be worse-off because they have fewer genes working; viruses increase the number of genes working.

One concern is that we may end up with thousands of vaccines to prevent every virus that can change our cells. Viruses can have different effects on different parts of the body. It may be possible to prevent a number of different cancers by immunizing against one viral agent. It is likely that HPV, for instance, has a role in cancers of the mucous membranes of the mouth.

Etiology

It is impossible to determine the initial cause for any specific cancer. However, with the help of molecular biological techniques, it is possible to characterize the mutations or chromosomal aberrations within a tumor, and rapid progress is being made in the field of predicting prognosis based on the spectrum of mutations in some cases. For example, upto half of all tumors have a defective p53 gene. This mutation is associated with poor prognosis, since those tumor cells are less likely to go into apoptosis or programmed cell death when damaged by therapy. Telomerase mutations remove additional barriers, extending the number of times a cell can divide. Other mutations enable the tumor to grow new blood vessels to provide more nutrients, or to metastasize, spreading to other parts of the body.

Cancer Stem Cells

A new way of looking at carcinogenesis comes from integrating the ideas of developmental biology into oncology. The cancer stem cell paradigm proposes that some or all cancers arise from transformation of adult stem cells. These cells persist as a subcomponent of the tumor and retain key stem cell properties. Furthermore, the relapse of cancer and the emergence of metastasis are also attributed to these cells. The cancer stem cell hypothesis does not contradict earlier concepts of carcinogenesis. It simply points to adult stem cells as the site where the process begins.

Clonal Evolution

Somatic Evolution in Cancer

While genetic and epigenetic alterations in tumor suppressor genes and oncogenes change the behavior of cells, those alterations, in the end, result in cancer through their effects on the population of neoplastic cells and their micro environment. Mutant cells in neoplasms compete for space and resources. Thus, a clone with a mutation in a tumor suppressor gene or oncogene will expand only in a neoplasm if that mutation gives the clone a competitive advantage over the other clones and normal cells in its micro environment. Thus, the process of carcinogenesis is formally a process of Darwinian evolution, known as somatic or clonal evolution. Furthermore, in light of the Darwinistic mechanisms of carcinogenesis, it has been theorized that the various forms of cancer can be categorized as pubertarial and gerontological. Anthropological research is currently being conducted on cancer as a natural evolutionary process through which natural selection destroys environmentally inferior phenotypes while supporting others. According to this theory, cancer comes in two separate types: from birth to the end of puberty (approximately age 20) teleologically inclined towards supportive group dynamics, and from mid-life to death (approximately age 40+) teleologically inclined away from overpopulative group dynamics.

Non-mainstream Theories

There are a number of theories of carcinogenesis and cancer treatment that fall outside the mainstream of scientific opinion, due to lack of scientific rationale, logic, or evidence base. These theories may be used to justify various alternative cancer treatments. They should be distinguished from those theories of carcinogenesis that have a logical basis within mainstream cancer biology, and from which conventionally-testable hypotheses can be made.

Teratology

Teratology is the study of abnormalities of physiological development. It is often thought of as the study of birth defects, but it is much broader than that, taking in other developmental stages, such as puberty; and other life forms, such as plants. The term stems from the Greek ôÝñ?ò (téras, genitive ôÝñ?ôïò- tératos), meaning *monster*, or *marvel* and ëüãïò - lógos, meaning *speech* or, more loosely, *the study of*.

Etymology

As early as the 17th century, Teratology referred to a discourse on prodigies and marvels, of anything so extraordinary as to seem abnormal. In the 19th century, it acquired a meaning more closely related to biological deformities, mostly in the field of botany. Currently, its most instrumental meaning is that of the medical study of teratogenesis, congenital malformations or grossly deformed individuals. Monster is a pejorative term for a grossly deformed individual.

Teratology as a medical term was popularized in the 1960s by Dr. David W. Smith of the University of Washington Medical School, one of the researchers who became known in 1973 for the discovery of Fetal alcohol syndrome. With greater understanding of the origins of birth defects, the field of teratology now overlaps with other fields of basic science, including developmental biology, embryology, and genetics.

In Humans and Other Mammals

Teratogenesis

Birth defects are known to occur in 3-5% of all newborns. They are the leading cause of infant mortality in the United States, accounting for more than 20% of all infant deaths. Seven to ten percent of all children will require extensive medical care to diagnose or treat a birth defect. Although significant progress has been made in identifying etiologic causes of some birth defects, approximately 65% have no known or identifiable cause.

It was previously believed that the mammalian embryo developed in the impervious uterus of the mother, protected from all extrinsic factors. However, after the thalidomide disaster of the 1960s, it became apparent and more accepted that the developing embryo could be highly vulnerable to certain environmental agents that have negligible or non-toxic effects to adult individuals.

Wilson's Six Principles

Along with this new awareness of the in utero vulnerability of the developing mammalian embryo came the development and refinement of *The Six Principles of Teratology* which are still applied today. These principles of teratology were put forth by Jim Wilson in 1959 and in his monograph *Environment and Birth Defects.* These principles guide the study and understanding of teratogenic agents and their effects on developing organisms:

1. Susceptibility to teratogenesis depends on the genotype of the conceptus and the manner in which this interacts with adverse environmental factors.
2. Susceptibility to teratogenesis varies with the developmental stage at the time of exposure to an adverse influence. There are critical periods of susceptibility to agents and organ systems affected by these agents.
3. Teratogenic agents act in specific ways on developing cells and tissues to initiate sequences of abnormal developmental events.
4. The access of adverse influences to developing tissues depends on the nature of the influence. Several factors affect the ability of a teratogen to contact a developing conceptus, such as the nature of the agent itself, route and degree of maternal exposure, rate of placental transfer and systemic absorption, and composition of the maternal and embryonic/foetal genotypes.
5. There are four manifestations of deviant development (Death, Malformation, Growth Retardation and Functional Defect).
6. Manifestations of deviant development increase in frequency and degree as dosage increases from the No Observable Adverse Effect Level (NOAEL) to a dose producing 100% Lethality (LD100).

Studies designed to test the teratogenic potential of environmental agents use animal model systems (e.g., rat, mouse, rabbit, dog, and monkey). Early teratologists exposed pregnant animals to environmental agents and observed the foetuses for gross visceral and skeletal abnormalities. While this is still part of the teratological evaluation procedures today, the field of Teratology is moving to a more molecular level, seeking the mechanism(s) of action by which these agents act. Genetically modified mice are commonly used for this purpose. In addition, pregnancy registries are large, prospective studies that monitor exposures in women received during their pregnancies and record the outcome of their births. These studies provide information about possible risks of medications or other exposures in human pregnancies.

Understanding how a teratogen causes its effect is not only important in preventing congenital abnormalities but also has the potential for developing new therapeutic drugs safe for use with pregnant women.

Teratology Education

It is estimated that 10% of all birth defects are caused by a prenatal exposure or teratogen. These exposures include, but are not limited to, medication or drug exposures, maternal infections and diseases, and environmental and occupational exposures. Teratogen-caused birth defects are potentially preventable. Studies have shown that nearly 50% of pregnant women have been exposed to at least one medication during gestation. An additional study found that of 200 individuals referred for genetic counseling for a teratogenic exposure, 52% were exposed to more than one potential teratogen.

Teratogenic Agents

A wide range of different chemicals and environmental factors are suspected or are known to be teratogenic in humans and in animals. A selected few include:

- Ionizing radiation: atomic weapons, radioiodine, radiation therapy
- Infections: cytomegalovirus, herpes virus, parvovirus B-19, rubella virus (German measles), syphilis, toxoplasmosis, Venezuelan equine encephalitis virus
- Metabolic imbalance: alcoholism, endemic cretinism, diabetes, folic acid deficiency, iodine deficiency, hyperthermia, phenylketonuria, rheumatic disease and congenital heart block, virilizing tumors
- Drugs and environmental chemicals: 13-cis-retinoic acid, isotretinoin (Accutane), temazepam (Restoril; Normisson), nitrazepam (Mogadon), nimetazepam (Ermin), aminopterin, androgenic hormones, busulfan, captopril, enalapril, chlorobiphenyls (PCBs), dioxin, coumarin, cyclophosphamide, diethylstilbestrol, diphenylhydantoin (Phenytoin, Dilantin, Epanutin), ethanol, ethidium bromide, etretinate, hexachlorobenzene, hexachlorophene, lithium, methimazole, organic mercury, penicillamine, tetracyclines, thalidomide, trimethadione, uranium, methoxyethyl ethers and valproic acid.

The status of some of the above substances (e.g. diphenylhydantoin) is subject to debate, and many other compounds are under varying degrees of suspicion. These include Agent Orange, nicotine, aspirin and other NSAIDs. Other compounds are known as severe teratogens based on veterinary work and animal studies, but aren't listed above because they have not been studied in humans, e.g. cyclopamine. Teratogenic effects also help to determine the pregnancy category assigned by regulatory authorities; in the United States, a pregnancy category of X, D, or C may be assigned if teratogenic effects (or other risks in pregnancy) are documented or cannot be excluded.

Exposure to teratogens can result in a wide range of structural abnormalities such as cleft lip, cleft palate, dysmelia, anencephaly, ventricular septal defect. In most cases, specific agents produce a specific teratogenic response.

In Plants

In botany, teratology investigates the theoretical implications of abnormal specimens. For example, the discovery of abnormal flowers— flowers with leaves instead of petals, or flowers with staminoid pistils—furnished important evidence for the "foliar theory", the theory that all flower parts are highly specialised leaves.

Chapter 12

Environmental Concepts

Pre-Industrial Toxicology

As humans sought to better understanding of natural compounds that were both beneficial and harmful to them, there was very little, if any, clear understanding of the fundamental-chemical nature of substances. That is, there was no connection between the extract or essence of a poisonous plant or animal and any one particular chemical that might cause toxicity. In fact, an awareness of chemistry in its modern form did not occur until around the mid to late 1600s. So it is ironic that Paracelsus, a physician from the sixteenth century and one of the early Fathers of Toxicology had no clear understanding of chemistry as we know it today. He, along with many others of that time, likely believed that all matter was composed of three primary bodies sulfur, salt, and mercury. Yet, Paracelsus also coined the now famous (or infamous) maxim of the newly emerging discipline of toxicology:

All substances are poisons, there is none which is not a poison. The right dose differentiates a poison from a remedy. (Paracelsus, 1493-1541)

This phrase and Paracelsus name are committed to memory by hundreds of new toxicology students each year and has become the motto of toxicology. Interestingly, if one takes Paracelsus at face value, it appears that in this quote he was referring to substances which served as potential remedies but could be poisonous if taken in high enough concentrations. Most of us are aware of the fact that overdosing can turn remedies to poisons, even with such apparently innocuous drugs such as aspirin and tylenol.

Another branch on the toxicology family tree that developed in the sixteenth century, along with the study of drugs and the use of chemicals in hunting and warfare, was occupational toxicology. As humans learned how to remove and exploit such materials as coal, and metals and other minerals, occupational exposures to these chemical substances (and chemicals produced incidentally) resulted. Scientists eventually recognized the linkages among illnesses and exposures to these compounds. Some of the first reports of occupational illness, or diseases caused by activities related to specific occupations, can be found in literature from the mid to late 1500s. Early occupational observations include the ill effects from lead mining and madness caused by mercury exposure. Later, in the 1700s, Bernadino Ramazzini is credited with bringing to light diseases of tradesmen, including silicosis in stone workers and lead poisoning. In the late 1700s, Sir Percival Potts made one of the more famous observations in toxicology, linking an occupational exposure (in this case, soot in chimney sweeps) to cancer of the scrotum.

The pre-Industrial Revolution developments in toxicology, that were primarily devoted to the study of such naturally-occurring toxicants as the polyaromatic compounds contained in soot and heavy metals, and such toxins as botulinum toxin produced by the bacterium *Clostridum botulinus.*

Toxicology and the Chemical and Industrial Revolution

The Chemical/Industrial Revolution of the mid-19th century released many naturally-occurring chemicals into the environment in unprecedented amounts. Also, it produced and released new substances unlike any that had existed in the natural world. With the production and use of these chemicals, and the need to protect humans from the toxic effects of industrial chemicals, toxicology eventually evolved to include its modern day branches: pharmacology, pesticide toxicology, general toxicology, and occupational toxicology.

Towards the mid-late 20th century, environmental toxicology was developed to specifically address the effects on both humans and wildlife of chemicals released into the environment. A notable difference among the branches of toxicology is that pharmacology, pesticides and even occupational toxicology primarily have focused on the effects of relatively high concentrations of single chemicals. This compares to the relatively low concentrations of several different chemicals or chemical mixtures that are relevant to environmental toxicology.

The chemicals considered by the earlier branches of toxicology were, and are, a known quantity. That is, the research was designed to address questions about specific, well-characterized chemicals, exposure conditions, and even concentration ranges rather than complex chemical mixtures. For example, pharmacologists might work with a particular active ingredient (e.g., salicylic acid or aspirin), and be confident about the route of exposure (oral) and the concentration or dose. This is seldom the case in environmental toxicology, and hazardous waste assessment and cleanup in particular, where chemicals often are present in mixtures, routes of exposure may vary (for example, from oral to dermal to inhalation). Significantly, exposure concentrations prove difficult to determine.

BRANCHES OF TOXICOLOGY SCIENCES

1. Clinical Toxicology: **(Hospital setting)**
   * Deal with emergencies such as overdoses, poisonings, attempted suicides.
   * Compound identification and quantification.
   * Sign and symptom management.
   * Emergency care—home poisoning.
   * Poison control.
2. Forensic Toxicology: **(Medical Examiner's office)**
   * Medical-legal aspects of poisonings.

- * Identification and quantification of poisons.
- * Establish relationship between tissue residual level and probable cause of death.

3. Industrial Toxicology:
   - * Estimation of worker safety based on 8 hr. work/day, 40 hr. work/week.
   - * Engineering of safety measures.
   - * Air sampling, worker sampling.
4. Environmental Toxicology:
   - * Effects of compounds on water, wildlife.
   - * Movement of chemicals in the environment—soil, air, water.
   - * Residual life of chemicals in the environment.
5. Biochemical and Molecular Toxicology:
   - * Determining mode of action of chemicals at the molecular level.
   - * Effect of chemicals on DNA, cancer genes.
6. Product Development Toxicology: **(Corporate setting)**
   - * Service and pre-clinical toxicology for product development.
   - * Evaluation of full toxic potential of chemicals destined for drug use.
   - * Establish safe doses for people.
7. Regulatory Toxicology: **(Industry and government setting)**
   - * Deciphers and analyzes toxicological data for risk estimation.
   - * Solvent vapour thresholds in industry.
   - * Safe level for human drugs.
   - * Safe level of heavy metals in water.
   - * Safe level of pesticides.

Performance Study of Coke Oven By-product Plant

In an integrated steel industry, wastewater generated from Coke Oven By-product Plant (COBP) is a major polluting stream containing toxic chemicals like phenol, cyanide and ammonia. In recent times, COBP effluent is treated by biochemical oxidation of cyanide, ammonia and phenol at separate effluent treatment plants, commonly known as bio chemical oxidation and dephenolisation plants (BOD plants). The treated effluent from BOD plant is recycled for use in quenching hot coke in coke-ovens. However, excess treated water has to be discharged into common outlet drain of steel plant. Almost all the integrated iron and steel units have installed BOD plants. Wastewater reduction in COBP involves reduction of pollution load in process wastewater and proper operation of BOD plant. Detailed study was conducted to evaluate the performance of BOD plants based on proper design, inlet concentration, type of treatment units, operation and maintenance and trained manpower.

Pollution Control in Coal-Based Thermal Power Plants

There are 81 coal-based thermal power plants comprising total installed capacity of 60263.5 MW. Thermal power plants are mainly responsible for air and water pollution besides problems related to solid waste (flyash) management. Out of 81 thermal power plants in the country, 42 plants comply with emission standards, 3 plants are closed and remaining 36 plants are in the process of installation/ augmentation of pollution control systems. As regards water pollution prevention measures, 49 plants comply with ash pond effluent standards, 3 plants are closed and 36 plants are in the process of installation/augmentation of ash ponds. The main factors affecting the performance of Electrostatic Precipitators are:

- Low design efficiency of pollution control equipments installed in various thermal power plants;
- Inefficient operation;
- High ash content in supplied coal; and,
- High resistivity of fly ash.

CETP for Sago Industries

Sago industries in Tamil Nadu (around 1000 units) are based on tapioca roots and mostly under small scale industries (SSI) sector. The treatment plants set up by individual units are not being properly operated. It is therefore proposed to set up a CETP for the sago industries located along Thirumanimuttar river. The wastewater from sago industries located along Thirumanimuttar riverbanks will be collected and treated at the CETP. Tamil Nadu Pollution Control Board has been requested to carry out the inventory of starch and sago industries in Salem town.

Common Chrome Recovery Plant

In Jajmau area of Kanpur, more than 300 tanneries of varying capacity are in operation. Many of the large and medium tanneries have installed the chromium recovery plant at the individual level. For small tanneries, CPCB has taken up a project to recommend cost-effective possibilities for installation of common chrome recovery plant.

Inventory and Pollution Control Measures in Lime Kiln Units

In order to prepare an inventory of lime kiln units, all the SPCBs/UTs were requested to provide the updated list of lime kiln units operating in their States/ UTs as well as the information regarding implementation of pollution control measures taken up by these units. The information is required to finalise inventory of lime kiln units being operated all over the country and to assess the magnitude of pollution load generated from these units.

Pollution Control In Small Scale Industries

The small scale industries (SSI) covering a wide spectrum of industries in small, tiny and cottage sector account for 40% of total industrial production in the country, 35% of total exports, manufactures more than 7,500 products and employ about 16.7 million person. However, unplanned, uncontrolled and haphazard growth of these industries have resulted in aggravating pollution problems. CPCB has undertaken an exercise to estimate the pollution load from industrial sources and to assess the status of pollution control in various sectors of small scale industries.

Human Risk Assessment in Asbestos Industries

Asbestos is a proven carcinogen and its exposure is associated with the development of lung cancer and mesothelioma. CPCB initiated human risk assessment studies in asbestos industries in collaboration with the Industrial Toxicology Research Centre, Lucknow. The study includes environmental monitoring of asbestos-based industries, detailed medical examinations of industrial workers, geno toxicological evaluation of asbestos exposed population and recommendations for preventive measures in industrial units.

Oil Drilling Units

Preliminary survey and inventorization of the oil-drilling units has been carried out in Upper Assam. The oilfields of Upper Assam, owned by Oil India Limited and Oil & Natural Gas Commission Limited, were visited to study the pollution aspects and also groundwater quality monitoring was conducted in the surroundings of injection wells.

Industrial survey statistics tells us that more than one-third – 36.3 percent – of the total value added by the raw materials through manufacture in the factory sector of the country comes from Maharashtra (23.66 percent) and Gujarat (12.64 percent). Easily, the two most industrialised states of India. Governments of both the States claim that they have created immense prosperity in the region. But statistics does not tell the real story of thousands of workers and farmers.

While driving down the Mumbai-Pune highway, one will witness the horrible truth of industrialisation. Hundreds of industrial units dealing with chemicals and fertilisers dump their sludge along the roadside. Chimneys emit gases that make breathing difficult. Industrial units never stop polluting, and people cannot stop working for them.

The CPCB (Central Pollution Control Board) has identified seventeen categories of industries (large and medium scale) as significantly polluting and the list includes highly air polluting industries such as integrated iron and steel, thermal power plants, copper/zinc/aluminium smelters, cement, oil refineries, petrochemicals, pesticides and fertiliser units. The state-wise distribution of these pre-1991 industries indicates that the states of Maharashtra, Uttar Pradesh, Gujarat, Andhra

Pradesh and Tamil Nadu have a large number of industries in these sectors. The category-wise distribution of these units reveals that sugar sector has the maximum number of industries, followed by pharmaceuticals, distillery, cement and fertiliser. It also indicates that agro-based and chemical industries have major shares of 47% and 37% of the total number of industries respectively. The status of pollution control as on 30 June 2000 is as follows: out of 1,551 industries, 1,324 have so far been provided the necessary pollution, 165 industries have been closed down and the remaining 62 industries are defaulters. It may be noted that in some of the key sectors such as iron and steel, 6 out of 8 units belong to the defaulters category in terms of having pollution control facilities to comply with the standards. On the other hand, cement, petrochemicals and oil refinery sectors do not have any defaulters. Small scale industries are a special feature of the Indian economy and play an important role in pollution. India has over 3 million small scale units accounting for over 40 percent of the total industrial output in the country. In general, Indian small scale industries lack pollution control mechanisms. While the larger industries are better organised to adopt pollution control measures, the small scale sector is poorly equipped (both financially and technically) to handle this problem. They have a very high aggregate pollution potential. Also, in many urban centres, industrial units are located in densely populated areas, thereby affecting a large number of people.

Toxicity

Toxicity is the degree to which a substance is able to damage an exposed organism. Toxicity can refer to the effect on a whole organism, such as an animal, bacterium, or plant, as well as the effect on a substructure of the organism, such as a cell (cytotoxicity) or an organ (organotoxicity), such as the liver (hepatotoxicity). By extension, the word may be metaphorically used to describe toxic effects on larger and more complex groups such as the family unit or society at large.

A central concept of toxicology is that effects are dose-dependent; even water can lead to water intoxication when taken in large enough doses, whereas for even a very toxic substance such as snake venom there is a dose below which there is no detectable toxic effect.

Types of Toxicity

There are generally three types of toxic entities chemical, biological, and physical.

- **Chemicals** include inorganic substances such as lead, mercury, asbestos, hydrofluoric acid, and chlorine gas, organic compounds such as methyl alcohol, most medications, and poisons from living things.
- **Biological** toxic entities include those bacteria and viruses that are able to induce disease in living organisms. Biological toxicity can be complicated to measure because the "threshold dose" may be a single organism. Theoretically one virus, bacterium or worm can reproduce to cause a serious

infection. However, in a host with an intact immune system the inherent toxicity of the organism is balanced by the host's ability to fight back; the effective toxicity is then a combination of both parts of the relationship. A similar situation is also present with other types of toxic agents.

- **Physically** toxic entities include things not usually thought of under the heading of "toxic" by many people: direct blows, conclussion, sound and vibration, heat and cold, non-ionizing electromagnetic radiation such as infrared and visible light, and ionizing radiation such as X-rays and alpha, beta, and gamma radiation.

Toxicity can be measured by the effects on the target (organism, organ, tissue or cell). Because individuals typically have different levels of response to the same dose of a toxin, a population-level measure of toxicity is often used which relates the probability of an outcome for a given individual in a population. One such measure is the LD_{50}. When such data does not exist, estimates are made by comparison to known similar toxic things, or to similar exposures in similar organisms. Then "safety factors" are added to account for uncertainties in data and evaluation processes. For example, if a dose of toxin is safe for a laboratory rat, one might assume that one-tenth of that dose would be safe for a human, allowing a safety factor of 10 to allow for interspecies differences between two mammals; if the data are from fish, one might use a factor of 100 to account for the greater difference between the two chordate classes (fish and mammals). Similarly, an extra protection factor may be used for individuals believed to be more susceptible to toxic effects such as in pregnancy or with certain diseases. Or, a newly synthesized and previously unstudied chemical that is believed to be very similar in effect to another compound could be assigned an additional protection factor of 10 to account for the possible differences in effects that are probably much smaller. Obviously, this approach is very approximate; but such protection factors are deliberately very conservative and the method has been found to be useful in a wide variety of applications.

Assessing all aspects of the toxicity of cancer-causing agents involves additional issues, since it is not certain if there is a minimal effective dose for carcinogens, or whether the risk is just too small to see. In addition, it is possible that a single cell transformed into a cancer cell is all it takes to develop the full effect.

It is more difficult to assess the toxicity of chemical mixtures than of single, pure chemicals because each component display its own toxicity and components may interact to produce enhanced or diminished effects. Common mixtures include gasoline, cigarette smoke, and industrial waste. Even more complex are situations with more than one type of toxic entity, such as the discharge from a malfunctioning sewage treatment plant, with both chemical and biological agents.

Global Classification of Toxicity

For substances to be regulated and handled appropriately they must be properly classified and labeled. Classification is determined by approved testing measures or calculations and have determined cut-off levels set by governments and scientists. While currently many countries have different regulations regarding the types of tests, amounts of tests and cut-off levels, the implementation of Global Harmonization will begin unifing these countries as early as 2008. Global Classification looks at three areas: Physical Hazards (explosions and pyrotechnics), Health Hazards and Environmental Hazards.

Health Hazards

Separates out the types of toxicities where substances may cause lethality to the entire body, lethality to specific organs, major/minor damage, or cause cancer wherein the cancer is what causes the damage and/or lethality.

Acute Toxicity

Acute toxicity looks at lethal effects following oral, dermal or inhalation exposure. It is broken into five categories of severity where Category 1 requires the least amount of exposure to be lethal and Category 5 requires the most exposure to be lethal.

1. Oral
   a. Category 1: LD50 = 5 mg/kg of bodyweight
   b. Category 2: LD50 = 50 mg/kg of bodyweight
   c. Category 3: LD50 = 300 mg/kg of bodyweight
   d. Category 4: LD50 = 2000 mg/kg of bodyweight
   e. Category 5: LD50 = 5000 mg/kg of bodyweight
   f. Substances beyond category 5 are not defined, though they cannot legally be defined as orally toxic.
2. Dermal
   a. Category 1: LD50 = 50 mg/kg of bodyweight
   b. Category 2: LD50 = 200 mg/kg of bodyweight
   c. Category 3: LD50 = 1,000 mg/kg of bodyweight
   d. Category 4: LD50 = 2,000 mg/kg of bodyweight
   e. Category 5: LD50 = 5,000 mg/kg of bodyweight
   f. Substances beyond category 5 are not defined, though they cannot legally be defined as dermally toxic.
3. Inhalation - Gases
   a. Category 1: LC50 = 100 ppmV
   b. Category 2: LC50 = 500 ppmV

c. Category 3: LC50 = 2,500 ppmV
d. Category 4: LC50 = 20,000 ppmV
e. Category 5: not defined though is expected to have an equivalent of category 5 oral and dermal

4. Inhalation – Vapors
   a. Category 1: LC50 = 0.5 mg/l
   b. Category 2: LC50 = 2.0 mg/l
   c. Category 3: LC50 = 10 mg/l
   d. Category 4: LC50 = 20 mg/l
   e. Category 5: not defined though is expected to have an equivalent of category 5 oral and dermal
5. Inhalation – Dust and Mists
   a. Category 1: LC50 = 0.05 mg/l
   b. Category 2: LC50 = 0.5 mg/l
   c. Category 3: LC50 = 1.0 mg/l
   d. Category 4: LC50 = 5 mg/l
   e. Category 5: not defined though is expected to have an equivalent of category 5 oral and dermal

Skin Corrosion and Irritation

Skin corrosion and irritation are determined through skin patch test analysis which looks at the severity of the damage done, when it is incurred, how long it remains, whether it is reversible and how many test subjects were affected.

Skin corrosion from a substance must penetrate through the epidermis into the dermis within four hours of application and must not reverse the damage within 14 days.

Skin irritation shows average damage (less severe than corrosion) within 72 hours of application or for three consecutive days after application within a 14-day period; or, inflammation that lasts for 14 days in two test subjects.

Mild skin irritation is minor damage (less severe than irritation) within 72 hours of application or for three consecutive days after application.

Eye Damage and Irritation

Serious eye damage involves tissue damage or decay of vision which does not fully reverse in 21 days.

Eye irritation involves changes to the eye which do not fully reverse within 21 days.

Respiratory or Skin Sensitization

Substances determined to cause the majority of test subjects either in breathing hypersensitivity through inhalation or allergic responses from dermal application.

Germ Cell Mutagenicity

Substances that can cause mutations to germ cells in an adult which can be passed to progeny.

Carcinogenicity

Substances that can induce cancer or the likeliness of cancer to occur.

Reproductive Toxicity

Adverse effects in sexual function and fertility in adults and/or development of the offspring. Where a substance can do either or both, though is not lethal to the adult.

Specific Target Organ Toxicity

Substances which can cause adverse effects to a specific organ through single or multiple exposures.

Aspiration Hazard

Direct/indirect ingestion or inhalation of a solid or liquid which can enter the lungs and cause sickness, damage or death.

Environmental Hazards

Environmental hazards tend to focus on degradability, bioaccumulation and aquatic toxicity.

Aquatic Toxicity

Aquatic toxicity testing submerges key indicator species of fish or crustacea to certain concentrations of a substance in their environment to determine the lethality level. Fishes are exposed for 96 hours while crustacea are exposed for 48 hours.

Acute Aquatic Toxicity Levels

Acute toxicity generally involves short-term exposure to adult species.

1. Category Acute 1: Fish (LC50) and/or crustacea (EC50) and/or aquatic plants = 1.0 mg/l
2. Category Acute 2: Fish (LC50) and/or crustacea (EC50) and/or aquatic plants >1 = 10 mg/l
3. Category Acute 3: Fish (LC50) and/or crustacea (EC50) and/or aquatic plants >10 = 100 mg/l

While GHS does not define toxicity past 100 mg/l, the EPA currently lists aquatic toxicity as "practically non-toxic" in concentrations greater than 100 ppm.

Chronic Aquatic Toxicity Levels

Chronic toxicity involves an analysis of long-term exposure, including life cycle stages. This method of analysis has less standardized testing and less data.

1. Category Chronic 1: Fish (LC50) and/or crustacea (EC50) and/or aquatic plants = 1.0 mg/l
2. Category Chronic 2: Fish (LC50) and/or crustacea (EC50) and/or aquatic plants >1 = 10 mg/l
3. Category Chronic 3: Fish (LC50) and/or crustacea (EC50) and/or aquatic plants >10 = 100 mg/l
4. Category Chronic 4: Substances which do not easily dissolve in water and have no acute toxicity values recorded.

Factors Influencing Toxicity

Toxicity of a substance can be affected by many different factors, such as the pathway of administration (whether the toxin is applied to the skin, ingested, inhaled, injected), the time of exposure (a brief encounter or long-term), the number of exposures (a single dose or multiple doses over time), the physical form of the toxin (solid, liquid, gas), the genetic makeup of an individual, an individual's overall health, and many others.

Acute exposure: A single exposure to a toxic substance which may result in severe biological harm or death; acute exposures are usually characterized as lasting no longer than a day.

Chronic exposure: Continuous exposure to a toxin over an extended period of time, often measured in months or years can cause irreversible side effects.

Chapter 13

Impact of Pollutant on Ecosystem

An ecosystem is a self-contained, dynamic system made of a population of species in its physical environment. This concept is used to study the complex interactions between the organisms - plants, animals, bacteria, and fungi - that make up the community. There are many different ways in which the community of organisms interacts.

There is the food chain where each organism is in a producer, consumer, predator, and prey relationship; there is the oxygen cycle and the water cycle that sustains the organisms. When an ecosystem gets polluted, the natural balance in the system is disturbed and this affects the organisms in different ways. It is important to know how a simple act like introducing sewage water or toxic waste into a lake can threaten several life species and plants in the area.

Pollution and the Ecosystem

Pollutants like oil, detergents, nitrogen and phosphate from fertilizers and lead can have a tremendous impact on the ecosystem, especially if the water gets polluted. In a lake, for example, it can wreak havoc on the ecological balance by stimulating plant growth and causing the death of fish due to suffocation resulting from lack of oxygen. The oxygen cycle will stop, and the polluted water will also affect the animals dependant on the lake water.

The Effect of Pollution on an Ecosystem

To understand how pollution impacts an ecosystem, an ecosystem is selected in neighbourhood. It could be a lake, a river or a sea shoreline. The different organisms in the ecosystem and how each organism is related to the other; which level are they in the food chain; how do they adapt to their environment.

One or two major organisms in the ecosystems are tracked then their behaviour, food habits, life cycle, breeding habits and there coping with the changes in the ecosystem is tracked. Though these changes are usually slow, one can track them over a period of time, say six months or one year. All the observations and findings are noted down, the specific details of each species that is being studied is taken into account.

TYPES OF ECOSYSTEM

MAJOR ECOSYSTEMS

Nevertheless, the ecosystem concept has proved very useful in ecology. One way in which it is applied is to describe the world's major different habitat types.

Terrestrial ecosystems include: Arctic and alpine ecosystems, dominated by cold areas and lacking trees; forest ecosystems, which can be subdivided into a whole range of types including tropical rainforests, mediterranean evergreen forests, temperate and boreal forests, and temperate deciduous forests; grasslands and savannahs; and deserts and semi-arid ecosystems. Freshwater ecosystems include lakes, rivers, and marshland. "Hybrid" terrestrial and freshwater systems include swamp forests and seasonal floodplains. Marine ecosystems comprise an enormous range, from coral reefs, mangroves, sea-grass beds, and other coastal and shallow water ecosystems to open-water ecosystems and the mysterious, little-known systems of the abyssal plains and trenches of the world's oceans. The term ecosystem may also be used to describe geographical areas which contain a wide range of habitat types which are linked by ecological processes. Thus the Serengeti-Seronera region in East Africa, one of the most spectacular wildlife areas in the world, is often regarded as a single ecosystem whose different constituent habitats include open grassy plains, woody savannah, dense thickets, and patches of forest, as well as rocky outcrops (kopjes), rivers, streams, and seasonal pools. Similarly, the most productive parts of the world's oceans have been divided into a series of Large Marine Ecosystems which contain many different habitats. Examples of Large Marine Ecosystems with widely different characteristics include the whole of the Black Sea, the Benguela Current system off the coast of south-western Africa, and the Gulf of Mexico.

NATURAL CHANGES IN ECOSYSTEM

The natural world is in a constant state of flux. Changes can be seen at all time-scales, from the shortest to the longest. Short-term changes, which are observable by people, are often cyclical and predictable: night and day, the monthly cycle of the tides, the annual change of the seasons, and the growth, reproduction, and death of individuals. Viewed at this level many ecosystems, when not disturbed by humans, appear superficially to be stable and unchanging, maintained in equilibrium by the "balance of nature".

It is becoming increasingly apparent that this is not the case. However, longer-term changes—those running over decades, centuries, millennia, and ultimately over tens of millions of years—are far less easy for us to track. The science of ecology itself is less than a century old—the merest blink of an eye in the history of most natural ecosystems. Moreover, it is evident that most of these longer-term changes are not regular and predictable at all.

Globally, climate is undoubtedly the most important factor in the short to medium term. On land, temperature, rainfall, and seasonality are the three factors which are most important in determining the distribution of ecosystems. Changes in any one of these can have a lasting effect. In recent geological time the most dramatic example of this is undoubtedly the series of Ice Ages which characterized much of the Pleistocene epoch. These protracted periods of global cooling had a major effect on the world's ecosystems, causing the spread of ice cover in temperate areas and the shrinking of moist forest habitats in parts of the tropics.

Climatic change with a wide geographical influence can also occur over shorter time scales. One of the most striking example is El Niño, a cold-water current which periodically sweeps across the Pacific. This has a major impact on marine ecosystems, for example, leading, to the death of reef corals in many parts of the Pacific, and the lowering of fisheries production in the Humboldt Current ecosystem off Chile and Peru. El Niños occur irregularly and vary in intensity and impact; it is rare for more than 20 years to pass without one and occasionally they may be separated by only 1 or 2 years. El Niños also affect terrestrial ecosystems through changing rainfall patterns, particularly in the Americas.

Local events can also have major effects on ecosystems: fires, floods, and landslides are all natural events which can have catastrophic local impact. This impact is not necessarily negative: many ecosystems actually require periodic disturbance to maintain themselves. Fire-climax ecosystems, in which periodic burning is an essential part of the growth cycle, are widespread in semi-arid areas such as much parts of Australia.

At longer time scales geological processes and evolution itself play a crucial role in changing the functioning of ecosystems. Continental Drift literally changes the face of the planet, creating new landscapes and destroying others, while evolution leads to new life-forms which may themselves create new ecosystems at the same time as leading to the extinction of other species and the loss or transformation of the ecosystems of which they were a part.

However, this is not to mean that there is no continuity in natural ecosystems. Many show great resilience and persistence through periods of many millions of years. Examples of ecosystems which have apparently remained stable for a very long time include those on the vast deep-sea plains, the Mediterranean-type ecosystems of southern Africa and western Australia, and some areas of tropical rainforest, such as those in parts of the south-east Asian mainland and in the mountains of eastern Africa.

HUMAN INFLUENCE ON ECOSYSTEM

All natural environments and ecosystems now have an unprecedented problem to deal with: humanity. Humans appear to be squeezing into a few short centuries, changes on a scale which would otherwise be expected over thousands or millions of years. The full impact of these remains to be guessed at. Major human impacts on ecosystems include the following:

A. HABITAT DESTRUCTION AND FRAGMENTATION

The most direct impact of humans on ecosystems is in their destruction or conversion. Clear-cutting (the cutting of all trees within a given forest area) will, obviously, destroy a forest ecosystem. Selective logging may also alter forest ecosystems in important ways. Fragmentation—the division of a once continuous ecosystem into a number of smaller patches—may disrupt ecological processes so that the remaining areas can no longer function as they once did.

B. CLIMATE CHANGE

It is now widely accepted that humanity's activities are contributing to global warming, chiefly through the accumulation of "greenhouse" gases in the atmosphere. The impact of this is likely to increase in the future. Climate change is a natural feature of the Earth. Previously, however, its effects were mitigated as ecosystems could effectively "migrate" by moving latitude or altitude as the climate changed. Today, so much of the world's land surface has been appropriated by people that in many cases there is no such place for the remaining natural or semi-natural ecosystems to migrate to.

C. POLLUTION

Contamination of the natural environment through a range of pollutants—herbicides, pesticides, fertilizers, industrial effluents, and human waste products—is one of the most pernicious forms of impact on the natural environment. Pollutants are often invisible, and the effects of air pollution and water pollution may not be immediately obvious, although they can be devastating in the long-run. The impact of acid rain on freshwater and forest ecosystems in much of northern and central Europe is a case in point.

D. INTRODUCED SPECIES

Human beings have been responsible either deliberately or accidentally for altering the distribution of a vast range of animal and plant species. This includes not only domesticated animals and cultivated plants but pests such as rats, mice, and many insects and fungi. Species which become naturalized may have a devastating impact, through predation and competition, on natural ecosystems, particularly on islands where native species have evolved in isolation. For instance, foxes, rabbits, cane toads, feral cats, and even buffaloes and camels have wreaked havoc in many ecosystems in Australia. Plants such as the South American shrub *Lantana* have invaded natural forests in many tropical and subtropical islands, causing major changes to these ecosystems, while the African water hyacinth *Eichhornia* has similarly disrupted freshwater ecosystems in many of the warmer parts of the world.

ECOSYSTEM OVER HARVEST

Removal of excessive number of animals or plants from a system can cause major ecological changes. The most important example of this at present is the over-fishing of the world's oceans. Depletion of the great majority of accessible fish stocks is undoubtedly a cause of major change, although its long-term impact is difficult to assess.

CONTROLLING HUMAN IMPACT ON ECOSYSTEM

Controlling changes in the world's ecosystems may be the major challenge facing human beings in the coming millennium. Solutions will have to be found at all scales, from the local to the global.

Protection of remaining natural ecosystems in national parks and other protected areas is crucial. However, this will not prevent areas being affected by factors such as climate change, or air- or water-borne pollutants. Moreover, as natural areas shrink in size they are likely to require more and more active management to maintain their ecological functions, for example, through control of exotic species, manipulation of water levels in wetlands, or periodic controlled burning in some forest habitats. Increased intervention of this sort will always be risky, as we still do not fully understand the working of most of the ecosystems.

Control of pollution and emission of greenhouse gases will require action at the global level, as will efforts to prevent further deterioration of marine fisheries through over-fishing. Ultimately, the solution lies in control of human population growth and in a far more restrained approach to our use of natural resources and expenditure of energy.

Chapter 14

Biodegradation, Biotransformation and Acid Rain

Biodegradation is the chemical breakdown of materials by a physiological environment. The term is often used in relation to ecology, waste management and environmental remediation (bioremediation). Organic material can be degraded aerobically with oxygen, or anaerobically without oxygen. A term related to biodegradation is biomineralisation, in which organic matter is converted into minerals. Biosurfactant, an extracellular surfactant secreted by microorganisms, enhances the biodegradation process.

Biodegradable matter is generally organic material such as plant and animal matter and other substances originating from living organisms, or artificial materials that are similar enough to plant and animal matter to be put to use by microorganisms. Some microorganisms have the astonishing, naturally occurring, microbial catabolic diversity to degrade, transform or accumulate a huge range of compounds including hydrocarbons (e.g. oil), polychlorinated biphenyls (PCBs), polyaromatic hydrocarbons (PAHs), pharmaceutical substances, radionuclides and metals. Major methodological breakthroughs in microbial biodegradation have enabled detailed genomic, metagenomic, proteomic, bioinformatic and other high-throughput analyses of environmentally relevant microorganisms providing unprecedented insights into key biodegradative pathways and the ability of microorganisms to adapt to changing environmental conditions.

BIOTRANSFORMATION

Biotransformation is the process whereby a substance is changed from one chemical to another (transformed) by a chemical reaction within the body. Metabolism or metabolic transformations are terms frequently used for the biotransformation process. However, metabolism is sometimes not specific for the transformation process but may include other phases of toxicokinetics.

Biotransformation is vital to survival in that it transforms absorbed nutrients (food, oxygen, etc.) into substances required for normal body functions. For some pharmaceuticals, it is a metabolite that is therapeutic and not the absorbed drug. For example, phenoxybenzamine (Dibenzyline®), a drug given to relieve hypertension, is biotransformed into a metabolite, which is the active agent. Biotransformation also serves as an important defense mechanism in that toxic xenobiotics and body wastes are converted into less harmful substances and substances that can be excreted from the body.

Toxicants that are lipophilic ('lipid-loving', dissolve easily in lipids), non-polar, and of low molecular weight are readily absorbed through the cell membranes of the skin, gastrointestinal (GI) tract, and lung. These same chemical and physical properties control the distribution of a chemical throughout the body and its penetration into tissue cells. Lipophilic toxicants are hard for the body to eliminate and can accumulate to hazardous levels. However, most lipophilic toxicants can be transformed into hydrophilic ('water-loving', dissolve easily in water) metabolites that are less likely to pass through membranes of critical cells. Hydrophilic chemicals are easier for the body to eliminate than lipophilic substances. Biotransformation is thus a key body defense mechanism.

Fortunately, the human body has a well-developed capacity to biotransform most xenobiotics as well as body wastes. An example of a body waste that must be eliminated is hemoglobin, the oxygen-carrying iron-protein complex in red blood cells. Hemoglobin is released during the normal destruction of red blood cells. Under normal conditions hemoglobin is initially biotransformed to bilirubin, one of a number of hemoglobin metabolites. Bilirubin is toxic to the brain of newborns and, if present in high concentrations, may cause irreversible brain injury. Biotransformation of the lipophilic bilirubin molecule in the liver results in the production of water-soluble (hydrophilic) metabolites excreted into bile and eliminated via the feces.

The biotransformation process is not perfect. When biotransformation results in metabolites of lower toxicity, the process is known as detoxification. In many cases, however, the metabolites are more toxic than the parent substance. This is known as bioactivation. Occasionally, biotransformation can produce an unusually reactive metabolite that may interact with cellular macromolecules (e.g., DNA). This can lead to very serious health effects, for example, cancer or birth defects. An example is the biotransformation of vinyl chloride to vinyl chloride epoxide, which covalently binds to DNA and RNA, a step leading to cancer of the liver.

Chemical Reactions

Chemical reactions are continually taking place in the body. They are a normal aspect of life, participating in the building up of new tissue, tearing down of old tissue, conversion of food to energy, disposal of waste materials, and elimination of toxic xenobiotics. Within the body is a magnificent assembly of chemical reactions, which is well-orchestrated and called upon as needed. Most of these chemical reactions occur at significant rates only because specific proteins, known as enzymes, are present to catalyze them, that is, accelerate the reaction. A catalyst is a substance that can accelerate a chemical reaction of another substance without itself undergoing a permanent chemical change.

Enzymes are the catalysts for nearly all biochemical reactions in the body. Without these enzymes, essential biotransformation reactions would take place slowly or not at all, causing major health problems. An example is the inability of persons that have phenylketonuria (PKU) to use the artificial sweetener, aspartame (in Equal®). Aspartame is basically phenylalanine, a natural constituent of most protein-containing foods. Some persons are born with a genetic condition in which the enzyme that can biotransform phenylalanine to tyrosine (another amino acid), is defective. As a result, phenylalanine can build up in the body and cause severe mental retardation. Babies are routinely checked at birth for PKU. If they have PKU, they must be given a special diet to restrict the intake of phenylalanine in infancy and childhood.

These enzymatic reactions are not always simple biochemical reactions. Some enzymes require the presence of co-factors or co-enzymes in addition to the substrate (the substance to be catalyzed) before their catalytic activity can be exerted. These co-factors exist as a normal component in most cells and are frequently involved in common reactions to convert nutrients into energy (vitamins are an example of co-factors). It is the drug or chemical transforming enzymes that hold the key to xenobiotic transformation. The relationship of substrate, enzyme, co-enzyme, and transformed product is illustrated below:

$$\text{substrate} \xrightarrow[\text{co-enzyme}]{\text{enzyme}} \text{transformed product}$$

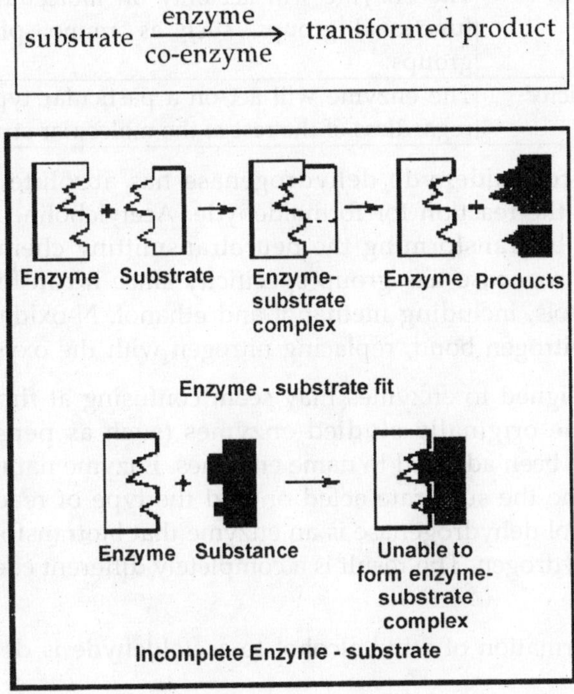

Figure 1

Most biotransforming enzymes are high molecular weight proteins, composed of chains of amino acids linked together by peptide bonds. A wide variety of biotransforming enzymes exist. Most enzymes will catalyze the reaction of only a few substrates, meaning that they have high "specificity". Specificity is a function of the enzyme's structure and its catalytic sites. While an enzyme may encounter many different chemicals, only those chemicals (substrates) that fit within the enzymes convoluted structure and spatial arrangement will be locked on and affected. This is sometimes referred to as the "lock-and-key" relationship. As shown in Figure 1, when a substrate fits into the enzyme's structure, an enzyme-substrate complex can be formed. This allows the enzyme to react with the substrate with the result that two different products are formed. If the substrate does not fit into the enzyme, no complex will be formed and thus no reaction can occur.

The array of enzymes range from those that have absolute specificity to those that have broad and overlapping specificity. In general, there are three main types of specificity:

Absolute Specificity	The enzyme will catalyze only one reaction.
Group Specificity	The enzyme will act only on molecules that have specific functional groups, such as amino, phosphate or methyl groups.
Linkage Specificity	The enzyme will act on a particular type of chemical bond regardless of the rest of the molecular structure.

For example, formaldehyde dehydrogenase has absolute specificity since it catalyzes only the reaction for formaldehyde. Acetylcholinesterase has absolute specificity for biotransforming the neurotransmitting chemical, acetylcholine. Alcohol dehydrogenase has group specificity since it can biotransform several different alcohols, including methanol and ethanol. N-oxidation can catalyze a reaction of a nitrogen bond, replacing nitrogen with the oxygen.

The names assigned to enzymes may seem confusing at first. However, except for some of the originally studied enzymes (such as pepsin and trypsin), a convention has been adopted to name enzymes. Enzyme names end in "ase" and usually combine the substrate acted on and the type of reaction catalyzed. For example, alcohol dehydrogenase is an enzyme that biotransforms alcohols by the removal of a hydrogen. The result is a completely different chemical, an aldehyde or ketone.

The biotransformation of ethyl alcohol to acetaldehyde is depicted below:

$$H-\underset{H}{\overset{H}{C}}-\underset{H}{\overset{H}{C}}-O-H \xrightleftharpoons[NAD]{ADH} H-\underset{H}{\overset{H}{C}}-\underset{H}{C}=O$$

ethyl alcohol acetaldehyde

ADH = alcohol dehydrogenase, a specific catalyzing enzyme; NAD = nicotinamide adenine dinucleotide, a common cellular reducing agent

It is clear that the transformation of a specific xenobiotic can be either beneficial or harmful—perhaps both depending on the dose and circumstances. A good example is the biotransformation of acetaminophen (Tylenol®), a commonly used drug to reduce pain and fever. When the prescribed doses are taken, the desired therapeutic response is observed with little or no toxicity. However, when excessive doses of acetaminophen are taken, hepatotoxicity can occur. This is because acetaminophen normally undergoes rapid biotransformation with the metabolites quickly eliminated in the urine and feces.

At high doses, the normal level of enzymes may be depleted and the acetaminophen is available to undergo reaction by an additional biosynthetic pathway, which produces a reactive metabolite that is toxic to the liver. For this reason, a user of Tylenol® is warned not to take the prescribed dose more frequently than every 4-6 hours and not to consume more than four doses within a 24-hour period. Biotransforming enzymes, like most other biochemicals, are available in a normal amount and in some situations can be "used up" at a rate that exceeds the body's ability to replenish them. This illustrates the frequently used phrase, the "Dose Makes the Poison."

Biotransformation reactions are categorized not only by the nature of their reactions, e.g., oxidation, but also by the normal sequence with which they tend to react with a xenobiotic. They are usually classified as Phase I and Phase II reactions. Phase I reactions are generally reactions which modify the chemical by adding a functional structure. This allows the substance to "fit" into the Phase II enzyme so that it can become conjugated (joined together) with another substance. Phase II reactions consist of those enzymatic reactions that conjugate the modified xenobiotic with another substance. The conjugated products are larger molecules than the substrate and generally polar in nature (water-soluble). Thus, they can be readily excreted from the body. Conjugated compounds also have poor ability to cross cell membranes.

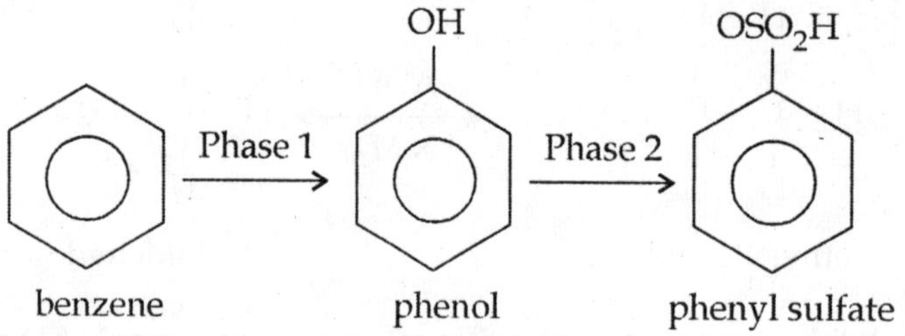

Figure 2

In some cases, the xenobiotic already has a functional group that can be conjugated and the xenobiotic can be biotransformed by a Phase II reaction without going through a Phase I reaction. A good example is phenol that can be directly conjugated into a metabolite that can then be excreted. The biotransformation of benzene requires both Phase I and Phase II reactions. As illustrated below, benzene is biotransformed initially to phenol by a Phase I reaction (oxidation). Phenol has the functional hydroxyl group that is then conjugated by a Phase II reaction (sulphation) to phenyl sulfate.

The major transformation reactions for xenobiotics are listed below:

Phase I	Phase II
Oxidation	Sulfate conjugation
Reduction	Glucuronide conjugation
Hydrolysis	Glutathione conjugation
Acetylation	Amino acid conjugation

Phase I Reactions

Phase I biotransformation reactions are simple reactions as compared to Phase II reactions. In Phase I reactions, a small polar group (containing both positive and negative charges) is either exposed on the toxicant or added to the toxicant. The three main Phase I reactions are oxidation, reduction and hydrolysis.

Oxidation is a chemical reaction in which a substrate loses electrons. There are a number of reactions that can achieve the removal of electrons from the substrate. Addition of oxygen was the first of these reactions discovered and thus the reaction was named oxidation. However, many of the oxidizing reactions do not involve oxygen. The simplest type of oxidation reaction is dehydrogenation, that is, the removal of hydrogen from the molecule. Another example of oxidation is electron-transfer that consists simply of the transfer of an electron from the substrate.

Examples of these types of oxidizing reactions are illustrated below:

Oxygenation

$$O{-}\underset{H}{\overset{H}{C}}{-}\underset{H}{\overset{H}{C}}{-} + \tfrac{1}{2}O_2 \xrightarrow{\text{Oxygenation}} {-}\underset{H}{\overset{H}{C}}{-}\underset{OH}{\overset{H}{C}}{-}$$

Dehydrogenation

$${-}\underset{H}{\overset{H}{C}}{-}\underset{H}{\overset{H}{C}}{-} \xrightarrow{\text{Dehydrogenation}} {-}\overset{H}{C}{=}\overset{H}{C}{-} + H_2$$

Electron Transfer

$$Fe^{3+} + Cu^{2+} \xrightarrow{\text{Electron Transfer}} Fe^{2+} + Cu^{3+}$$

The specific oxidizing reactions and oxidizing enzymes are numerous. Most of the reactions are self-evident from the name of the reaction or enzyme involved. Listed below are several of these oxidizing reactions:

- alcohol dehydrogenation
- aldehyde dehydrogenation
- alkyl/acyclic hydroxylation
- aromatic hydroxylation
- deamination
- desulfuration
- N-dealkylation
- N-hydroxylation
- N-oxidation
- O-dealkylation
- sulphoxidation

Reduction is a chemical reaction in which the substrate gains electrons. Reductions are most likely to occur with xenobiotics in which oxygen content is low. Reductions can occur across nitrogen-nitrogen double bonds (azo reduction) or on nitro groups (NO_2). Frequently, the resulting amino compounds are oxidized forming toxic metabolites. Some chemicals such as carbon tetrachloride can be reduced to free radicals, which are quite reactive with biological tissues. Thus, reduction reactions frequently result in activation of a xenobiotic rather than detoxification. An example of a reduction reaction in which the nitro group is reduced is illustrated below:

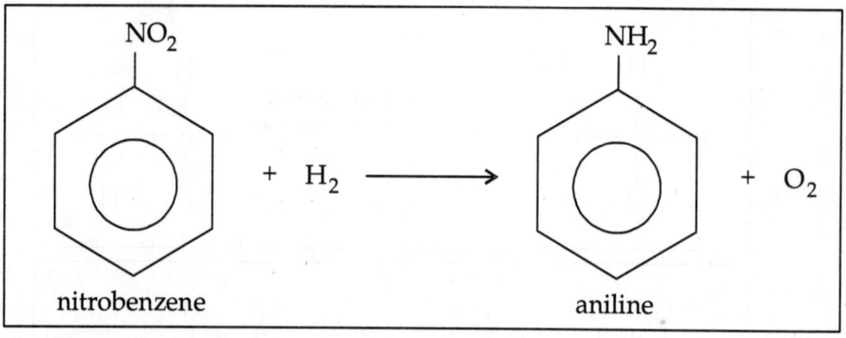

There are fewer specific reduction reactions than oxidizing reactions. The nature of these reactions is also self-evident from their name. Listed below are several of the reducing reactions.

- azo reduction
- dehalogenation
- disulfide reduction
- nitro reduction
- N-oxide reduction
- sulfoxide reduction

Hydrolysis is a chemical reaction in which the addition of water splits the toxicant into two fragments or smaller molecules. The hydroxyl group (OH^-) is incorporated into one fragment and the hydrogen atom is incorporated into the other. Larger chemicals such as esters, amines, hydrazines, and carbamates are generally biotransformed by hydrolysis. The example of the biotransformation of procaine (local anesthetic) which is hydrolyzed to two smaller chemicals is illustrated below:

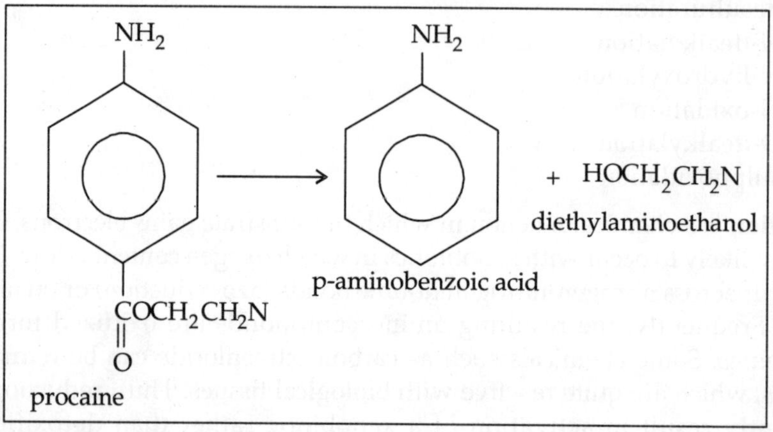

Toxicants that have undergone Phase I biotransformation are converted to metabolites that are sufficiently ionized, or hydrophilic, to be either eliminated from the body without further biotransformation or converted to an intermediate metabolite that is ready for Phase II biotransformation. The intermediates from Phase I transformations may be pharmacologically more effective and, in many cases, more toxic than the parent xenobiotic.

Phase II Reactions

A xenobiotic that has undergone a Phase I reaction is now a new intermediate metabolite that contains a reactive chemical group, e.g., hydroxyl (-OH), amino (-NH_2), and carboxyl (-COOH). Many of these intermediate metabolites do not possess sufficient hydrophilicity to permit elimination from the body. These metabolites must undergo additional biotransformation as a Phase II reaction.

Phase II reactions are conjugation reactions, that is, a molecule normally present in the body is added to the reactive site of the Phase I metabolite. The result is a conjugated metabolite that is more water-soluble than the original xenobiotic or Phase I metabolite. Usually the Phase II metabolite is quite hydrophilic and can be readily eliminated from the body. The primary Phase II reactions are:

- glucuronide conjugation - most important reaction
- sulfate conjugation - important reaction
- acetylation
- amino acid conjugation
- glutathione conjugation
- methylation

Glucuronide conjugation is one of the most important and common Phase II reactions. One of the most popular molecules added directly to the toxicant or its phase I metabolite is glucuronic acid, a molecule derived from glucose, a common carbohydrate (sugar) that is the primary source of energy for cells. The sites of glucuronidation reactions are substrates having an oxygen, nitrogen, or sulfur bond. This includes a wide array of xenobiotics as well as endogenous substances, such as bilirubin, steroid hormones and thyroid hormones. Glucuronidation is a high-capacity pathway for xenobiotic conjugation. Glucuronide conjugation usually decreases toxicity, although there are some notable exceptions, for example, the production of carcinogenic substances. The glucuronide conjugates are generally quite hydrophilic and are excreted by the kidney or bile, depending on the size of the conjugate. The glucuronide conjugation of aniline is illustrated below:

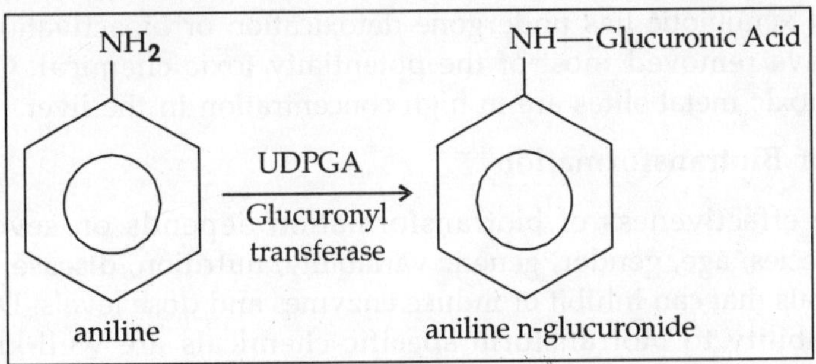

Sulfate conjugation is another important Phase II reaction that occurs with many xenobiotics. In general, sulfation decreases the toxicity of xenobiotics. Unlike glucuronic acid conjugates that are often eliminated in the bile, the highly polar sulfate conjugates are readily secreted in the urine. In general, sulfation is a low-capacity pathway for xenobiotic conjugation. Often glucuronidation or sulfation can conjugate the same xenobiotics.

Biotransformation Sites

Biotransforming enzymes are widely distributed throughout the body. However, the liver is the primary biotransforming organ due to its large size and high concentration of biotransforming enzymes. The kidneys and lungs are next with 10-30% of the liver's capacity. A low capacity exists in the skin, intestines, testes, and placenta. Since the liver is the primary site for biotransformation, it is also potentially quite vulnerable to the toxic action of a xenobiotic that is activated to a more toxic compound.

Within the liver cell, the primary subcellular components that contain the transforming enzymes are the microsomes (small vesicles) of the endoplasmic reticulum and the soluble fraction of the cytoplasm (cytosol). The mitochondria, nuclei, and lysosomes contain a small level of transforming activity.

Microsomal enzymes are associated with most Phase I reactions. Glucuronidation enzymes, however, are contained in microsomes. Cytosolic enzymes are non-membrane-bound and occur free within the cytoplasm. They are generally associated with Phase II reactions, although some oxidation and reduction enzymes are contained in the cytosol. The most important enzyme system involved in Phase I reactions is the cytochrome P-450 enzyme system. This system is frequently referred to as the "mixed function oxidase (MFO)" system. It is found in microsomes and is responsible for oxidation reactions of a wide array of chemicals.

The fact that the liver biotransforms most xenobiotics and that it receives blood directly from the gastrointestinal tract renders it particularly susceptible to damage by ingested toxicants. Blood leaving the gastrointestinal tract does not directly flow into the general circulatory system. Instead, it flows into the liver first via the portal vein. This is known as the "first pass" phenomena. Blood leaving the liver is eventually distributed to all other areas of the body; however, much of the absorbed xenobiotic has undergone detoxication or bioactivation. Thus, the liver may have removed most of the potentially toxic chemical. On the other hand, some toxic metabolites are in high concentration in the liver.

Modifiers of Biotransformation

The relative effectiveness of biotransformation depends on several factors, including species, age, gender, genetic variability, nutrition, disease, exposure to other chemicals that can inhibit or induce enzymes and dose levels. Differences in species capability to biotransform specific chemicals are well-known. Such differences are normally the basis for selective toxicity, used to develop chemicals as effective as pesticides but are relatively safe in humans. For example, malathion in mammals is biotransformed by hydrolysis to relatively safe metabolites, but in insects, it is oxidized to malaoxon, which is lethal to insects.

Safety testing of pharmaceuticals, environmental and occupational substances is conducted with laboratory animals. Often, differences between animal and human biotransformation are not known at the time of initial laboratory testing since information is lacking in humans. Humans have a higher capacity for glutamine conjugation than laboratory rodents. Otherwise, the types of enzymes and biotransforming reactions are basically comparable. For this reason, determination of biotransformation of drugs and other chemicals using laboratory animals is an accepted procedure in safety testing.

Age may affect the efficiency of biotransformation. In general, human fetuses and neonates (newborns) have limited abilities for xenobiotic biotransformations. This is due to inherent deficiencies in many, but not all, of the enzymes responsible for catalyzing Phase I and Phase II biotransformations. While the capacity for biotransformation fluctuates with age in adolescents, by early adulthood the enzyme activities have essentially stabilized. Biotransformation capability is also decreased in the aged. Gender may influence the efficiency of biotransformation for specific xenobiotics. This is usually limited to hormone-related differences in the oxidizing cytochrome P-450 enzymes.

Genetic variability in biotransforming capability accounts for most of the large variation among humans. The Phase II acetylation reaction, in particular, is influenced by genetic differences in humans. Some persons are rapid and some are slow acetylators. The most serious drug-related toxicity occurs in the slow acetylators, often referred to as "slow metabolizers". With slow acetylators, acetylation is so slow that blood or tissue levels of certain drugs (or Phase I metabolites) exceeds their toxic threshold.

Poor nutrition can have a detrimental effect on biotransforming ability. This is related to inadequate levels of proteins, vitamins, and essential metals. These deficiencies can decrease the ability to synthesize biotransforming enzymes. Many diseases can impair an individual's capacity to biotransform xenobiotics. A good example, is hepatitis (a liver disease), which is well-known to reduce hepatic biotransformation to less than half normal capacity.

Enzyme inhibition and enzyme induction can be caused by prior or simultaneous exposure to xenobiotics. In some situations exposure to a substance will inhibit the biotransformation capacity for another chemical due to inhibition of specific enzymes. A major mechanism for the inhibition is competition between the two substances for the available oxidizing or conjugating enzymes, that is, the presence of one substance uses up the enzyme that is needed to metabolize the second substance.

Enzyme induction is a situation where prior exposure to certain environmental chemicals and drugs results in an enhanced capability for biotransforming a xenobiotic. The prior exposures stimulate the body to increase the production of some enzymes. This increased level of enzyme activity results in increased biotransformation of a chemical subsequently absorbed. Examples of enzyme inducers are alcohol, isoniazid, polycyclic halogenated aromatic hydrocarbons (e.g., dioxin), phenobarbital, and cigarette smoke. The most commonly induced enzyme reactions involve the cytochrome P-450 enzymes.

Dose level can affect the nature of the biotransformation. In certain situations, the biotransformation may be quite different at high doses versus that seen at low dose levels. This contributes to the existence of a dose threshold for toxicity. The mechanism that causes this dose-related difference in biotransformation usually can be explained by the existence of different biotransformation pathways. At low doses, a xenobiotic may follow a biotransformation pathway that detoxifies the substance. However, if the amount of xenobiotic exceeds the specific enzyme capacity, the biotransformation pathway is "saturated". In that case, it is possible that the level of parent toxin builds up. In other cases, the xenobiotic may enter a different biotransformation pathway that may result in the production of a toxic metabolite.

An example of a dose-related difference in biotransformation occurs with acetaminophen (Tylenol®). At normal doses, approximately 96% of acetaminophen is biotransformed to non-toxic metabolites by sulfate and glucuronide conjugation. At the normal dose, about 4% of the acetaminophen is oxidized to a toxic metabolite; however, that toxic metabolite is conjugated with glutathione and excreted. With 7-10 times the recommended therapeutic level, the sulphate and glucuronide conjugation pathways become saturated and more of the toxic metabolite is formed. In addition, the glutathione in the liver may also be depleted so that the toxic metabolite is not detoxified and eliminated. It can react with liver proteins and cause fatal liver damage.

BIOMAGNIFICATION

Biomagnification, also known as bioamplification or biological magnification, is the increase in concentration of a substance, such as the pesticide DDT, that occurs in a food chain as a consequence of:

- Food chain energetics
- Low (or non-existent) rate of excretion/degradation of the substance.

Although sometimes used interchangeably with 'bioaccumulation', an important distinction is drawn between the two, and with bioconcentration, it is also important to distinguish between sustainable development and over-exploitation in biomagnification.

- **Bioaccumulation** occurs *within* a trophic level, and is the increase in concentration of a substance in an individual's tissues due to uptake from food and sediments in an aquatic milieu.
- **Bioconcentration** is defined as occurring when uptake from the water is greater than excretion.

Thus, bioconcentration and bioaccumulation occur within an organism, and biomagnification occurs across trophic (food chain) levels.

Lipid-soluble (lipophilic) substances cannot be excreted in urine, a water-based medium, and so accumulate in fatty tissues of an organism if the organism lacks enzymes to degrade them. When eaten by another organism, fats are absorbed in the gut, carrying the substance, which then accumulates in the fats of the predator. Since at each level of the food chain there is a lot of energy loss, a predator must consume many prey, including all of their lipophilic substances.

For example, though mercury is only present in small amounts in seawater, it is absorbed by algae (generally as methylmercury). It is efficiently absorbed, but only very slowly excreted by organisms. Bioaccumulation and biomagnification result in buildup in the adipose tissue of successive trophic levels: zooplankton, small nekton, larger fish etc. Anything which eats these fish also consumes the higher level of mercury the fish have accumulated. This process explains why predatory fish such as swordfish and sharks or birds like osprey and eagles have higher concentrations of mercury in their tissue than could be accounted for by direct exposure alone. For example, herring contains mercury at approximately 0.01 ppm and shark contains mercury at greater than 1 ppm.

Current Status

In a review of a large number of studies, it was concluded that although biomagnification is probably more limited in occurrence than previously thought, there is good evidence that DDT, DDE, PCBs, toxaphene, and the organic forms of mercury and arsenic do biomagnify in nature. For other contaminants, bioconcentration and bioaccumulation account for their high concentrations in organism tissues. The above studies refer to aquatic systems. In terrestrial systems, direct uptake by higher trophic levels must be much less, occurring via the lungs.

This concept of the biomagnification does not mean that we need not be concerned about synthetic organic contaminants and metal elements because they will become diluted. Bioaccumulation and bioconcentration result in these substances remaining in the organisms and not being diluted to non-threatening concentrations. The success of top predatory-bird recovery (bald eagles, peregrine falcons) in North America following the ban on DDT use in agriculture is testament to the importance of biomagnification.

Substances that Biomagnify

There are two main groups of substances that biomagnify. Both are lipophilic and not easily degraded. Novel organic substances are not easily degraded because organisms lack previous exposure and have thus not evolved specific detoxification and excretion mechanisms, as there has been no selection pressure from them. These substances are consequently known as 'persistent organic pollutants' or POPs.

Metals are not degradable because they are elements. Organisms, particularly those subject to naturally high levels of exposure to metals, have mechanisms to sequester and excrete metals. Problems arise when organisms are exposed to higher concentrations than usual, which they cannot excrete rapidly enough to prevent damage. These metals are transferred in an organic form.

Novel Organic Substances

Persistent Organic Pollutant:

- DDT
- PCBs
- Toxaphene
- Monomethylmercury

Inorganic Substances:

- Arsenic
- Cadmium
- Mercury

ACID RAIN

"Acid rain" is a popular term referring to the deposition of wet (rain, snow, sleet, fog and cloudwater, dew) and dry (acidifying particles and gases) acidic components. A more accurate term is "acid deposition". Distilled water, which contains no carbon dioxide, has a neutral pH of 7. Liquids with a pH less than 7 are acidic, and those with a pH greater than 7 are bases. "Clean" or unpolluted rain has a slightly acidic pH of about 5.2, because carbon dioxide and water in the air react together to form carbonic acid, a weak acid (pH 5.6 in distilled water), but unpolluted rain also contains other chemicals.

$$H_2O\ (l) + CO_2\ (g) \rightarrow H_2CO_3\ (aq.)$$

Carbonic acid then can ionize in water forming low concentrations of hydronium and carbonate ions:

$$2\ H_2O\ (l) + H_2CO_3\ (aq.) \rightleftharpoons CO_3^{2-}\ (aq.) + 2\ H_3O^+\ (aq.)$$

Emissions of chemicals leading to acidification:

The most important gas which leads to acidification is sulfur dioxide. Emissions of nitrogen oxides which are oxidized to form nitric acid are of increasing importance due to stricter controls on emissions of sulfur containing compounds. 70 Tg(S) per year in the form of SO_2 comes from fossil fuel combustion and industry, 2.8 Tg(S) from wildfires and 7-8 Tg(S) per year from volcanoes.

Natural Phenomena

The principal natural phenomena that contribute acid-producing gases to the atmosphere are emissions from volcanoes and those from biological processes that occur on the land, in wetlands, and in the oceans. The major biological source of sulfur containing compounds is dimethyl sulfide.

Acidic deposits have been detected in glacial ice thousands of years old in remote parts of the globe.

The principal cause of acid rain is sulfur and nitrogen compounds from human sources, such as electricity generation, factories, and motor vehicles. Coal power plants are one of the most polluting. The gases can be carried hundreds of kilometres in the atmosphere before they are converted to acids and deposited. In the past, factories had short funnels to let out smoke, but this caused many problems locally; thus, factories now have taller smoke funnels. However, dispersal from these taller stacks causes pollutants to be carried farther, causing widespread ecological damage.

Chemical Processes

Gas phase Chemistry

In the gas phase, sulfur dioxide is oxidized by reaction with the hydroxyl radical via an intermolecular reaction:

$$SO_2 + OH\cdot \rightarrow HOSO_2\cdot$$

which is followed by:

$$HOSO_2\cdot + O_2 \rightarrow HO_2\cdot + SO_3$$

In the presence of water, sulfur trioxide (SO_3) is converted rapidly to sulfuric acid:

$$SO_3\,(g) + H_2O\,(l) \rightarrow H_2SO_4\,(l)$$

Nitric acid is formed by the reaction of OH with nitrogen dioxide:

$$NO_2 + OH\cdot \rightarrow HNO_3$$

Chemistry in Cloud Droplets

When clouds are present, the loss rate of SO_2 is faster than can be explained by gas phase chemistry alone. This is due to reactions in the liquid water droplets.

Hydrolysis

Sulfur dioxide dissolves in water and then, like carbon dioxide, hydrolyses in a series of equilibrium reactions:

$$SO_2\,(g) + H_2O \rightleftharpoons SO_2 \cdot H_2O$$

$$SO_2 \cdot H_2O \rightleftharpoons H^+ + HSO_3^-$$

$$HSO_3^- \rightleftharpoons H^+ + SO_3^{2-}$$

Oxidation

There are a large number of aqueous reactions that oxidize sulfur from S(IV) to S(VI), leading to the formation of sulfuric acid. The most important oxidation reactions are with ozone, hydrogen peroxide and oxygen (reactions with oxygen are catalyzed by iron and manganese in the cloud droplets).

Acid Deposition

Wet Deposition

Wet deposition of acids occurs when any form of precipitation (rain, snow, etc.) removes acids from the atmosphere and delivers it to the Earth's surface. This can result from the deposition of acids produced in the raindrops or by the precipitation removing the acids either in clouds or below clouds. Wet removal of both gases and aerosols are both of importance for wet deposition.

Dry Deposition

Acid deposition also occurs via dry deposition in the absence of precipitation. This can be responsible for as much as 20 to 60% of total acid deposition. This occurs when particles and gases stick to the ground, plants or other surfaces.

Acid rain has been shown to have adverse impacts on forests, freshwaters and soils, killing insect and aquatic life-forms as well as causing damage to buildings and having impacts on human health.

Surface Waters and Aquatic Animals

Both the lower pH and higher aluminium concentrations in surface water that occur as a result of acid rain can cause damage to fish and other aquatic animals. At pHs lower than 5 most fish eggs will not hatch and lower pHs can kill adult fish. As lakes and rivers become more acidic, biodiversity is reduced. Acid rain has eliminated insect life and some fish species, including the brook trout in some lakes, streams, and creeks in geographically sensitive areas, such as the Adirondack Mountains of the United States. However, the extent to which acid rain contributes directly or indirectly via run-off from the catchment to lake and river acidity (i.e., depending on characteristics of the surrounding watershed) is variable.

Soils

Soil biology and chemistry can be seriously damaged by acid rain. Some microbes are unable to tolerate changes to low pHs and are killed. The enzymes of these microbes are denatured (changed in shape so they no longer function) by the acid. The hydronium ions of acid rain also mobilize toxins such as aluminium, and leach away essential nutrients and minerals such as magnesium.

$$2\ H^+\ (aq.) + Mg^{2+}\ (clay) \rightleftharpoons 2\ H^+\ (clay) + Mg^{2+}\ (aq.)$$

Soil chemistry can be dramatically changed when base cations, such as calcium and magnesium, are leached by acid rain thereby affecting sensitive species, such as sugar maple (Acer saccharum).

Forests and Other Vegetation

Adverse effects may be indirectly related to acid rain, like the acid's effects on soil or high concentration of gaseous precursors to acid rain. High altitude forests are especially vulnerable as they are often surrounded by clouds and fog which are more acidic than rain.

Other plants can also be damaged by acid rain but the effect on food crops is minimized by the application of lime and fertilizers to replace lost nutrients. In cultivated areas, limestone may also be added to increase the ability of the soil to keep the pH stable, but this tactic is largely unusable in the case of wilderness lands. When calcium is leached from the needles of red spruce, these trees become less cold tolerant and exhibit winter injury and even death.

Human Health

Scientists have suggested direct links to human health. Fine particles, a large fraction of which are formed from the same gases as acid rain (sulfur dioxide and nitrogen dioxide), have been shown to cause illness and premature deaths such as cancer and other diseases.

Acid rain can also cause damage to certain building materials and historical monuments. This results when the sulfuric acid in the rain chemically reacts with the calcium compounds in the stones (limestone, sandstone, marble and granite) to create gypsum, which then flakes off.

$$CaCO_3\ (s) + H_2SO_4\ (aq.) \rightleftharpoons CaSO_4\ (aq.) + CO_2\ (g) + H_2O\ (l)$$

This result is also commonly seen on old gravestones where the acid rain can cause the inscription to become completely illegible. Acid rain also causes an increased rate of oxidation for iron. Visibility is also reduced by sulfate and nitrate aerosols and particles in the atmosphere.

Affected Areas

Particularly badly affected places around the globe include most of Europe (particularly, Scandinavia with many lakes with acidic water containing no life and many trees dead), many parts of the United States (States like New York are very badly affected) and South-Western Canada. Other affected areas include the South-Eastern coast of China and Taiwan.

Potential Problem Areas in the Future

Places like much of South Asia (Indonesia, Malaysia and Thailand), Western-South Africa (the country), Southern India and Sri Lanka and even West Africa (countries like Ghana, Togo and Nigeria) could all be prone to acidic rainfall in the future.

Prevention Methods

Technical Solutions

In the United States, many coal-burning power plants use Flue gas desulfurization (FGD) to remove sulfur-containing gases from their stack gases. An example of FGD is the wet scrubber which is commonly used in the U.S. and many other countries. A wet scrubber is basically a reaction tower equipped with a fan that extracts hot smoke stack gases from a power plant into the tower. Lime or limestone in slurry form is also injected into the tower to mix with the stack gases and combine with the sulfur dioxide present. The calcium carbonate of the limestone produces pH-neutral calcium sulfate that is physically removed from the scrubber. That is, the scrubber turns sulfur pollution into industrial sulfates.

In some areas, the sulfates are sold to chemical companies as gypsum when the purity of calcium sulfate is high. In others, they are placed in landfill. However, the effects of acid rain can last for generations, as the effects of pH level change can stimulate the continued leaching of undesirable chemicals into otherwise pristine water sources, killing off vulnerable insect and fish species and blocking efforts to restore native life.

Automobile emissions control reduces emissions of nitrogen oxides from motor vehicles.

International Treaties

A number of international treaties on the long-range transport of atmospheric pollutants have been agreed, e.g., Sulphur Emissions Reduction Protocol under the Convention on Long-Range Transboundary Air Pollution.

Emission Trading

In this regulatory scheme, every current polluting facility is given or may purchase on an open market an emissions allowance for each unit of a designated pollutant it emits. Operators can then install pollution control equipment, and sell portions of their emissions allowance they no longer need for their own operations, thereby recovering some of the capital cost of their investment in such equipment. The intention is to give operators economic incentives to install pollution controls.

There are many forms of acid rain that are seen around the world. In parts of the world where there is wet weather, there is acid rain, acid snow, and acid fog. In parts of the world where there is dry weather, there is acid gas and acid dust. All of the lakes and streams in the world are normally slightly acidic. Heavy rainstorms or melting snow can cause the acidity in lakes and in streams to increase.

The Effect of Acid Rain on Sea Life

Acid rain is very harmful to the environment. Acid rain damages everything over a period of time because it makes the living things in the environment die. Acid rain affects the life in the water as well as the life on land. It is almost worse in water than on land because the fish that are in the water need the water to breathe. When the water gets polluted, then the fish get sick and end up dying.

All rain water contains some level of acidity. Acidity is measured by pH, which stands for potential of hydrogen. The pH scale measures the amount of acid in a substance. pH is measured on a scale from 0-14, with 7 being neutral. The lower the number is on the pH scale, the more acidic that substance is. Normal rain water has a pH of 5.6. When the pH level of rain water goes below 5.6, it is considered acid rain.

All of the sea life will die when the water that they swim in gets to be too acidic. For example, all fish will die when the water goes below a pH of 4.5. Most of the frogs and insects that live around the water will also die when the water reaches a pH of 4.5. With a pH of 5.5, all of the bottom-dwelling bacterial decomposers, animals that eat the remains of the food that other animals don't want, will begin to die. When these decomposers die, they leave the un decomposed food on the bottom of the water. This pollutes the water by making the water dirty for all of the fish to swim in. All fresh water shrimp die when there gets to be a pH of 6.0. Aquatic plants will grow the best when the water is at pH between 7.0 and 9.2. If acid rain gets to be more of a problem, then all of the sea life will eventually be gone.

Some of the lakes that were once acidic are recovering, but many more are not recovering. Of the 202 lakes that were chosen to be studied in the early 1980s only 33% of them have become less acidic.

The Effect of Acid Rain on the Forests of the World

Trees are also harmed by acid rain. Scientists say that acid rain damages the waxy outer coating that protects the leaves. When this happens, it allows the acid to seep into the tree. Instead of water changing from a liquid to a gas inside the leaves, gas is taking the place of the water. This prevents the plant from taking in carbon dioxide to perform photosynthesis, and the plant will eventually die. Acid rain, acid fog, and acid vapour also damage forests by damaging the surface of the leaves and needles. This makes it harder for the trees to withstand the cold and will cause the trees to die. Acid rain also harms the soil that the trees are growing in by taking most of the valuable nutrients away from the soil. Acid rain also leaves a lot of aluminium in the soil, which can be harmful to the trees that grow there.

The atmosphere deposits a lot of toxic metals into the forests because acid rain contains metal. Some of these metals are lead, zinc, copper, chromium, and aluminium. When there is acid rain, the rain releases these metals. This is believed to stunt the growth of many trees and plants. This also stunts the growth of mosses, algae, nitrogen-fixing bacteria, and fungi that are needed to help the forest to grow. Forests need these because they eat the harmful things that will kill the trees, such as bad bacteria. Acid rain hurts trees because they cannot grow any more.

The Effect of Acid Rain on the Air and Health

Acid rain affects us in many different ways. One major way is our health. Breathing and lung problems in children and adults who have asthma and in children have been linked to acid air pollution. Everything that we eat, drink, and breathe has at one time come in contact with acid deposits. This could threaten our health by making us become sick. The following health problems occur each year in the U.S. and Canada due to acid rain:

550 premature deaths,

1,520 emergency room visits,

210,070 asthma symptom days.

If acid rain became a little less of a problem, then there would be many health problems that could be avoided.

Effect of Acid Rain to Non-Living Things

Acid rain can also damage non-living things such as buildings and statues. It can decay building materials and paints. Worst of all, it can damage non-replaceable buildings, statues, and sculptures that are part of our nation's memories that we want to last for a very long time.

Acid rain is caused by:

Acid rain is mainly caused by these substances that are being released into the air:

•

Carbon dioxide: Carbon dioxide is released by burning coal, oil, and natural gas. If you inhale carbon dioxide, then since it is toxic, it can cause breathing problems, unconsciousness, and other serious health problems. **Carbon monoxide:** Carbon monoxide is released by burning gasoline, oil, and wood. When carbon monoxide enters your body, it goes into the bloodstream. When this happens, it will slow down the delivery of oxygen to the rest of the body, causing dizziness, headaches, and fatigue.

Chlorofluorocarbons (CFCs): CFCs are the chemicals that are used in industry, refrigeration, air-conditioning systems, and consumer products. Whenever CFCs are released into the air, they reduce the stratospheric ozone layer. The stratospheric ozone layer protects the Earth's surface from the harmful rays of the sun.

Hazardous air pollutants (HAPs): HAPs are released into the air by sources such as chemical plants, dry cleaners, printing plants, and motor vehicles (cars, trucks, buses, and planes). HAPs can cause serious health problems like cancer, birth defects, nervous system problems, and deaths that are all due to people accidentally letting them go into the air.

Lead: Lead is released by house and car paint as well as the manufacturing of lead batteries, fishing lures, certain parts of bullets, some ceramic ware, water pipes, and fixtures. In young children, lead can cause nervous system damage and learning problems.

Nitrogen oxides: Nitrogen oxides are released into the air by burning fuels such as gasoline and coal. When nitrogen oxides combine with Volatile Organic Compounds (VOCs), they can cause breathing difficulty in people who have asthma, coughs in children, and general illness in respiratory system.

Ozone: Ozone is released by motor vehicles, industries, burning coal, gasoline, and other fossil fuels, and in the chemicals that are in hairspray and paints. When ozone is close to the ground (ground level ozone) it can cause chest pain, irritated respiratory tract, or persistent cough, can make you unable to take deep breaths, and can make you more likely to get lung infections.

Particulate matter (PM): PM, little particles of pollution, is released by cars, trucks, and buses that are burning diesel fuel, fertilizers, pesticides, road construction, steel making, mining, and turning on fire places and wood stoves. When PMs mix with air particles and get breathed in by something, they get stuck in the lung tissue. There they can cause increased respiratory disease and lung damage.

Sulfur dioxides: Sulfur dioxides are released by burning coal, paper production, and melting metal. Sulfur dioxide can harm vegetation, harm metals, and cause lung problems, which include breathing problems and permanent lung damage.

Volatile organic compounds (VOCs): VOCs are released into the air by burning gasoline, wood, coal, or natural gas, solvents, paints, glues, and other products that are used at work or at home.

There are a lot of similarities in all of these pollutants. Most of the pollutants are from automobiles. Automobiles release harmful smoke into the air, which causes acid rain. Coal, oil, and gasoline are also some of the most common causes of all of the pollutants. If people reduce the amount of these things that they release into the air, then there will be less pollutants. Some of the most common health problems are breathing problems, nervous system problems, and lung problems.

Table : Air Pollutants

Air Pollutant	% that mobile sources contribute to acid rain	% that other sources contribute to acid rain
Volatile organic compound	37%	63%
Nitrogen oxide	49%	51%
Carbon monoxide	81%	19%
Particulate matter	27%	73%

The table above shows that the biggest air pollutant that mobile sources contribute to acid rain is carbon monoxide. Of all of the carbon monoxide releases that contribute to acid rain, 81% of them come from mobile sources. The biggest other source is particulate matter, little particles of pollution that are released into the air by cars, trucks, and buses that are burning diesel fuel, fertilizers, pesticides, road construction, steel making, mining, and turning on fire places and wood stoves. Seventy three percentage of the non-mobile sources that contribute to acid rain are caused by the release of particulate matter. The Table above shows how much mobile and other sources of pollution can make acid rain more of a problem. Seeing that carbon monoxide and particulate matter are the leading sources of pollution, by cutting down on these, acid rain will not be as much of a problem.

What can be Done?

There are many ways by which people can stop pollution. One major way is to reduce the amount of trips that one takes in your car. Another way that a lot of our pollution is caused is by creating electrical energy. When electricity is created, fuels are usually burned, and this causes the pollution, which causes acid rain. The generation of electric power produces more pollution than any other industry. Burning coal and other fossil fuels causes most of our pollution. This is why in some places around the world, acid rain is monitored very closely. In 1998, data shows that by using electricity, the pollution that comes with it was responsible for 67% of the sulfur dioxide emissions that caused acid rain that year. Every time that you turn on the lights, causes the pollution and that causes acid rain. Even doing little things that you may think don't cause pollution sometimes really do. Some things that you can do to make acid rain less of a problem are:

In Our Home

- Only run the dishwasher with a full load.
- Only run the washing machine with a full load.
- Turn off the lights in empty rooms or when you will be away from home.
- Turn off the hot water tank when you will be gone for a long period of time.
- Turn down the heat at night and when you will not be at home for the night.
- Don't use your air conditioner very much.
- Install fluorescent light bulbs instead of incandescent light bulbs.
- Try to reduce, reuse, and recycle as often as you can.
- Try not to burn a fire as often as you usually do.

In the Yard

- Keep the pool cover on the pool whenever you are not using it.

Transportation

- When you are going to work, you could walk, ride your bike, or take a bus.
- Car-pool to a place with someone else.
- For alternate fuels, try ethanol, propane, or natural gas.
- Take the train or a bus for long trips.
- Limit the amount of long trips you take in your car.
- Make sure that your vehicle's air conditioning system isn't leaking.
- Try not to overflow the gas tank.
- Make sure that you are travelling at high speeds only when you need to.

Acid rain hurts many things. Some things that are being hurt by acid rain are trees, animals, and most of all, sea life. People can help stop acid rain by not polluting the air. When the chemicals in the air turn into a gas and evaporate, they mix with the water vapour, and cause acid rain. Even kids can help prevent this pollution by using less electricity and using transportation that is friendly to the environment.

How Acid Rain Works?

On looking through the Appalachian Mountains, one will spot stands of dead and weakened trees. If one lives in a city, one might notice worn stone buildings, streaks on car roof or corroded metal railings and statues. The effects of acid rain appears nearly everywhere one goes, but with media and public attention turned to the more ominous prospect of global warming, acid rain has fallen by the wayside.

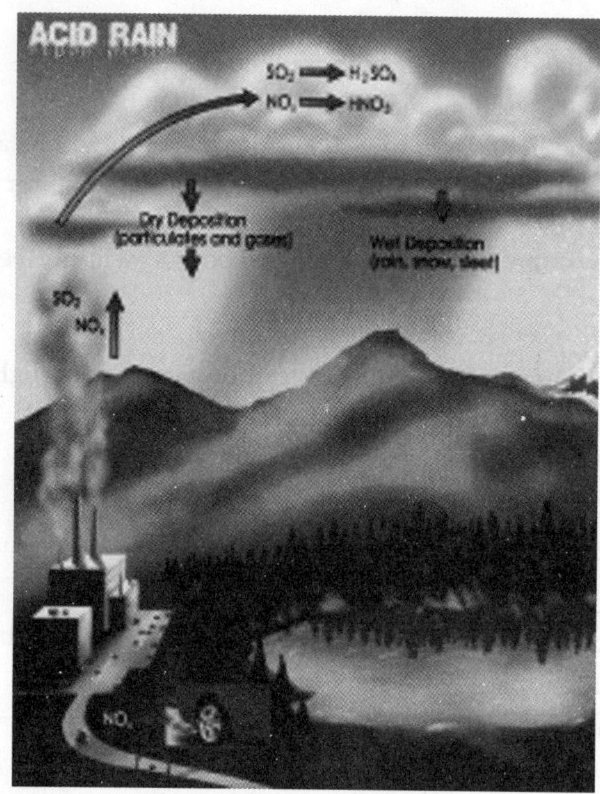

Emissions of sulfur dioxide and nitrogen oxides react with water vapor in the atmosphere to create sulfuric and nitric acids.

Acid rain occurs mostly in the Northern Hemisphere — the more industrialized, dirtier half of the globe. Winds can sweep up emissions from high smoke stacks and carry pollutants far from their original sources, crossing state lines and national borders in the process. Acid rain may not have the complete global range of greenhouse gases, but it is a transboundary, and therefore, international issue.

Acid rain, also known as acid deposition, is caused by emissions of sulfur dioxide (SO_2) and nitrogen oxides (NO) from power plants, cars and factories. Natural sources like volcanoes, forest fires and lightning strikes also add to the man-made pollution. SO_2 and NO become acids when they enter the atmosphere and react with water vapour. The resulting sulfuric and nitric acids can fall as wet or dry depositions. Wet deposition is precipitation: acid rain, snow, sleet or fog. Dry deposition falls as acidic particulates or gases.

The pH of Acid Rain

Scientists express the acidity of acid rain using the pH scale. The scale defines a solution's acidity, neutrality or alkalinity based on its concentration of hydrogen ions. Acids have a high concentration of hydrogen ions and a low pH. The scale ranges from 0 to 14, with pure water at a neutral 7.0. Most water, however, is not exactly pure. Even clean, normal rain has a pH of about 5.6. This is because it reacts with carbon dioxide in the atmosphere and forms mildly acidic carbonic acid before it becomes rain.

The pH scale is a measure of acidity and alkalinity. Acid rain has a pH of 5.0 or less.

Most acid deposition ranges from pH 4.3 to 5.0 — somewhere between the acidity of orange juice and black coffee. But comparing acid rain to safe, natural acids can be misleading. Even at its weakest, acid rain wrecks ecosystems by stunting sensitive plants and killing delicate aquatic eggs.

Programs that monitor acid rain analyze hydrogen content to determine pH. They also measure atmospheric concentrations of nitric acid, nitrate, sulfur dioxide, sulfate and ammonium. Monitoring acid deposition helps determine critical loads, or the amount of pollutants an ecosystem can support before damage. Accurate critical loads help in setting effective targets for SO_2 and NO reductions.

Surface Waters

Surface waters and their fragile ecosystems are perhaps the most famous victims of acid rain. Most of the precipitation that enters a lake, river, stream or marsh must first pass over and seep through soil. All soil has a buffering capacity, or ability to resist changes in acidity and alkalinity. The soil's buffering capacity determines a water body's acidity. If the capacity is low, or has reached its limit, acid rain can pass through unneutralized.

Most life is comfortable at a near-neutral pH – stray too far from pH 7.0, and delicate organisms begin to die. Plankton and invertebrates are sensitive to changes in acidity and die first. At pH 5.0, fish eggs degrade and young cannot develop. Adult fishes and frogs can sometimes tolerate acidities as low as pH 4.0, but they starve as their weaker food sources die out. When acid rain disrupts the food chain, biodiversity decreases.

Nitrogen deposition from acid rain also damages coastal waters and estuaries. Nitrogen-rich water supports massive algae growth and algal blooms. Bacteria decompose the dead algae, flourish themselves and soak up the water's available oxygen. Fish, shellfish, sea grass beds and coral reefs die in the algae-choked, oxygen-depleted waters. Scientists estimate that 10 percent to 45 percent of human-produced nitrogen that winds up in coastal waters comes from atmospheric deposition.

Most acidic bodies of water do not look polluted. As decaying organic matter settles, acidified water can appear clear and blue. Some species, like rushes and moss, even thrive in acidic conditions. But the greenery and clear waters belie an unwholesome environment. Diversity drops and species left without predators often grow disturbingly large.

The Effects of Acid Rain on Forests

Forests rely on their soil's buffering capacity to protect them from acid rain. Acidic waters draw out soil toxins like aluminium. Trees take in the poisonous substances, and runoff dumps it in lakes, rivers and streams. Acid rain also dissolves helpful minerals and nutrients like calcium, magnesium and potassium before trees can absorb them. Acid rain rarely kills a forest outright but instead stunts its growth through years of soil degradation. Nutrient deprivation and exposure to toxins make trees more likely to topple in storms or die in cold weather.

Even trees in well-buffered soil can weaken in harsh acid fog. High-elevation forests soak in acidic clouds, which strip leaves of nutrients and break down trees' ability to resist cold. The bald peaks of the Appalachian Mountains tell of the poisonous effect of acid rain on high-elevation forests.

Materials and Finishes

Acid rain has the unsettling ability to erase and obliterate stone and metal, the most durable of materials. Old buildings, monuments and tombstones bear the smooth signs of acidic corrosion and deterioration. Acid deposition speeds up natural weathering caused by rain, sun, snow and wind.

Acid rain also mars automotive paint. The auto industry considers acid deposition one type of corrosive **environmental fallout**, along with tree sap, pollen and bird droppings. Acid markings leave irregular, etched shapes on horizontal surfaces. Repainting is the only way to fix a car finish disfigured by acid rain.

Acid Rain and the Taj Mahal

Damage from acid rain mars some of the world's finest cultural monuments. Emissions reductions, however, have helped slow the rate of damage in North America and Europe. India's Taj Mahal has not fared as well. The mausoleum built by the Mughal Emperor Shah Jahan for his beloved wife Mumtaz Mahal is losing its white luster and turning a sickly pale shade. Scientists blame pollution from local foundries and a nearby oil refinery.

Health

Since acid rain can kill aquatic animals, weaken trees and dissolve stone, it seems like it could also scald or burn humans. But it doesn't affect people in the same way as it does fish or plants. Acid rain feels the same as regular rain – it's even safe to swim in an acidic lake. But the sulfate and nitrate particulates of dry deposition can cause asthma, bronchitis and heart problems. The NO in acid deposition also reacts with volatile organic compounds (VOCs) to form ground-level ozone. Ozone or smog aggravates and weakens the respiratory system.

Reducing Acid Rain

Acid rain has existed since the first factories of the Industrial Revolution began spitting out toxic emissions. An English scientist, Robert Angus Smith, coined the term "acid rain" in 1872 when he wrote of its corroding touch on buildings and deadly effect on plants. But acid rain did not become a government-monitored environmental problem until more than a century later. Scientists had by then determined that acid rain was a transboundary rather than a local concern. In 1980, the Acid Deposition Act launched a 10-year study on acid rain under the direction of the National Acidic Precipitation Assessment Program (NAPAP) to monitor sites around the country.

In 1990, armed with the NAPAP's study, Congress changed the existing Clean Air Act to include acid rain. The new Title IV amendment of the Clean Air Act called for SO_2 and NO reductions. The Acid Rain Program (ARP) was formed in 1995 to bring Title IV into effect.

The ARP places limits on the power industry to reduce annual emissions of SO_2 and NO. The ARP uses a cap and trade program to cut SO_2 emissions. After setting a cap, the ARP distributes allowances to power plant units. Units are only allowed to produce as much SO_2 as they have credit for. If they reduce emissions faster than the ARP requires, they can bank allowances for future use or sell them to other plants. The final 2010 cap will be 8.95 million tons allowed per year, a remarkable 50 percent less than power plant emissions from 1980.

NUCLEAR WINTER

Nuclear winter is a term that describes the predicted climatic effects of nuclear war. Severely cold weather and reduced sunlight for a period of months or years would be caused by detonating large numbers of nuclear weapons, especially over flammable targets such as cities, where large amounts of smoke and soot would be injected into the Earth's stratosphere. The term has also been applied to one of the after-effects of a comet or asteroid impact, also sometimes termed an impact winter, or of a supervolcano eruption, known as a volcanic winter.

Mechanism

The nuclear winter scenario predicts that the huge fires caused by nuclear explosions (particularly from burning urban areas) would loft massive amounts of dark smoke and aerosol particles from the fires into the upper troposphere/stratosphere. At 10-15 kilometers (6-9 miles) above the Earth's surface, the absorption of sunlight would further heat the smoke, lifting it into the stratosphere where the smoke would persist for years, with no rain to wash it out. This would block out much of the sun's light from reaching the surface, causing surface temperatures to drop drastically.

Consequences

Climatic Effects

Ozone Depletion

A 2008 study published in the Proceedings of the National Academy of Science found that nuclear weapons exchange between Pakistan and India using their current arsenals could create a near-global ozone hole, triggering human health problems and wreaking environmental havoc for at least a decade. The computer-modeling study looked at a nuclear war between the two countries involving 50 Hiroshima-sized nuclear devices on each side, producing massive urban fires and lofting as much as five million metric tons of soot about 50 miles (80 km) into the stratosphere. The soot would absorb enough solar radiation to heat surrounding gases, setting in motion a series of chemical reactions that would break down the stratospheric ozone layer protecting Earth from harmful ultraviolet radiation.

Column ozone losses could exceed 20% globally, 25-45% at mid-latitudes, and 50-70% at northern high latitudes persisting for 5 years, with substantial losses continuing for 5 additional years. Column ozone amounts would remain near or below 220 Dobson units at all altitudes even after three years, constituting an extra-tropical "ozone hole". Human health ailments like cataracts and skin cancer, as well as damage to plants, animals and ecosystems at mid-latitudes would likely rise sharply as ozone levels decreased and allowed more harmful UV light to reach Earth, according to the PNAS study. This study demonstrates that a small-scale, regional nuclear conflict is capable of triggering ozone losses even larger than losses that were predicted in the 1980s following a full-scale nuclear war. The missing piece back then was that the models at the time could not account for the rise of the smoke plume and consequent heating of the stratosphere.

Recent Modeling

Based on new work published in 2007 and 2008 by some of the pioneers of nuclear winter research who worked on the original studies, several things can now be said about this topic.

New Science:

- A minor nuclear war (such as between India and Pakistan or in the Middle East), with each country using 50 Hiroshima-sized atom bombs as airbursts on urban areas, could produce climate change unprecedented in recorded human history. This is only 0.03% of the explosive power of the current global arsenal.
- This same scenario would produce global ozone depletion, because the heating of the stratosphere would enhance the chemical reactions that destroy ozone.
- A nuclear war between the United States and Russia today could produce nuclear winter, with temperatures plunging below freezing in the summer in major agricultural regions, threatening the food supply for most of the planet.
- The climatic effects of the smoke from burning cities and industrial areas would last for several years, much longer than previously thought. New climate model simulations, that have the capability of including the entire atmosphere and oceans, show that the smoke would be lofted by solar heating to the upper stratosphere, where it would remain for years.

New Modeling

The climatic effects of smoke from fires started by nuclear war depend on the amount of smoke. For 50 nuclear weapons dropped on two countries, on the targets that would produce the maximum amount of smoke, about 5 megatons (Tg) of black smoke would be produced, accounting for the amount emitted from the fires and the amount immediately washed out in rain. As the smoke is lofted into the stratosphere, it would be transported around the world by the prevailing winds.

Two scenarios of war between the two Super Powers who still maintain large nuclear arsenals, the United States and Russia were calculated. In one scenario, 50 Tg of black smoke would be produced and in another, 150 Tg of black smoke would be produced. The number of nuclear weapons required to produce this much smoke depends on the targets, but there are enough weapons in the current arsenals to produce either amount. In fact, there are only so many targets, once they are all hit by weapons, additional weapons would not produce much more smoke at all.

Even after the current nuclear weapons reduction treaty between these super powers is played out in 2012, with each having about 2,000 weapons, 150 Tg of smoke could still be produced. These new results were made possible by the use of a state-of-the-art general circulation model of the climate.

For the first time a complete calculation of not only atmospheric but also oceanic circulation was conducted, including the entire atmosphere from the surface up through the troposphere, stratosphere, and mesosphere, to an elevation of 80 kilometers (50 miles). Previous calculations had not been run for the 10-year simulations here, and had not allowed the smoke to be lofted into the upper stratosphere, where it would persist for many years. The climate response to the above scenarios was calculated.

Compared to the global warming observed for the past century, all three scenarios show massive cooling. Compared to the climate change for the Northern Hemisphere for the past 1,000 years, the famous hockey stick diagram, the climate change from any of these scenarios is unprecedented.

Compared to climate change for the past millennium, even the 5 Tg case (a war between India and Pakistan) would plunge the planet into temperatures colder than the Little Ice Age (approximately 1600-1850). This would essentially effect agriculture and would be severely threatened. Larger amounts of smoke would produce larger climate changes, and for the 150 Tg case produce a true nuclear winter, making agriculture impossible for years. In both cases, new climate model simulations show that the effects would last for more than a decade.

RECYCLING

Recycling involves processing used materials into new products in order to prevent waste of potentially useful materials, reduce the consumption of fresh raw materials, reduce energy usage, reduce air pollution (from incineration) and water pollution (from landfilling) by reducing the need for "conventional" waste disposal, and lower greenhouse gas emissions as compared to virgin production. Recycling is a key component of modern waste management and is the third component of the "Reduce, Reuse, Recycle" waste hierarchy.

Recyclable materials include many kinds of glass, paper, metal, plastic, textiles, and electronics. Although similar in effect, the composting or other reuse of biodegradable waste – such as food or garden waste – is not typically considered recycling. Materials to be recycled are either brought to a collection center or picked up from the curbside, then sorted, cleaned, and reprocessed into new materials bound for manufacturing.

In a strict sense, recycling of a material would produce a fresh supply of the same material, for example, used office paper to more office paper, or used foamed polystyrene to more polystyrene. However, this is often difficult or too expensive (compared with producing the same product from raw materials or other sources), so "recycling" of many products or materials involves their reuse in producing different materials (e.g., cardboard) instead. Another form of recycling is the salvage of certain materials from complex products, either due to their intrinsic value (e.g., lead from car batteries, or gold from computer components), or due to their hazardous nature (e.g., removal and reuse of mercury from various items).

Critics dispute the net economic and environmental benefits of recycling over its costs. Specifically, critics argue that the costs and energy used in collection and transportation detract from (and possibly outweigh) the costs and energy saved in the production process; also that the jobs produced by the recycling industry can be a poor trade for the jobs lost in logging, mining, and other industries associated with virgin production; and that materials such as paper pulp can only be recycled a few times before material degradation prevents further recycling. Proponents of recycling counter each of these claims, and the validity of arguments from both sides has led to enduring controversy.

History

Early Recycling

Recycling has been a common practice for most of human history, with recorded advocates as far back as Plato in 400 BC. During periods when resources were scarce, archaeological studies of ancient waste dumps show less household waste (such as ash, broken tools and pottery)—implying more waste was being recycled in the absence of new material.

In pre-industrial times, there is evidence of scrap bronze and other metals being collected in Europe and melted down for perpetual reuse. In Britain, dust and ash from wood and coal fires was collected by 'dustmen' and down-cycled as a base material used in brick-making. The main driver for these types of recycling was the economic advantage of obtaining recycled feedstock instead of acquiring virgin material, as well as a lack of public waste removal in ever more densely populated areas. In 1813, Benjamin Law developed the process of turning rags into 'shoddy' and 'mungo' wool in Batley, Yorkshire. This material combined recycled fibres with virgin wool. The West Yorkshire shoddy industry in towns such as Batley and Dewsbury, lasted from the early 19s to at least the First World War.

Resource shortages caused by the World Wars, and other such world-changing occurrences greatly encouraged recycling. Massive government promotion campaigns were carried out in World War II in every country involved in the war, urging citizens to donate metals and conserve fibre, as a matter of significant patriotic importance. Resource conservation programs established during the war were continued in some countries without an abundance of natural resources, such as Japan, after the war ended.

Legislation

Supply

In order for a recycling program to work, having a large, stable supply of recyclable material is crucial. Three legislative options have been used to create such a supply: mandatory recycling collection, container deposit legislation, and refuse bans. Mandatory collection laws set recycling targets for cities to aim for, usually in the form that a certain percentage of a material must be diverted from the city's waste stream by a target date. The city is then responsible for working to meet this target.

Container deposit legislation involves offering a refund for the return of certain containers, typically glass, plastic, and metal. When a product in such a container is purchased, a small surcharge is added to the price. This surcharge can be reclaimed by the consumer if the container is returned to a collection point. These programs have been very successful, often resulting in an 80% recycling rate. Despite such good results, the shift in collection costs from local government to industry and consumers has created strong opposition to the creation of such programs in some areas.

A third method of increase supply of recyclates is to ban the disposal of certain materials as waste, often including used oil, old batteries, tires and garden waste. One aim of this method is to create a viable economy for proper disposal of banned products. Care must be taken that enough of these recycling services exist, or such bans simply lead to increased illegal dumping.

Government-mandated Demand

Legislation has also been used to increase and maintain a demand for recycled materials. Four methods of such legislation exist: minimum recycled content mandates, utilization rates, procurement policies, recycled product labeling.

Both minimum recycled content mandates and utilization rates increase demand directly by forcing manufacturers to include recycling in their operations. Content mandates specify that a certain percentage of a new product must consist of recycled material. Utilization rates are a more flexible option: industries are permitted to meet the recycling targets at any point of their operation or even contract recycling out in exchange for tradeable credits. Opponents to both of these methods point to the large increase in reporting requirements they impose, and claim that they rob industry of necessary flexibility.

Governments have used their own purchasing power to increase recycling demand through what are called "procurement policies". These policies are either "set-asides", which earmark a certain amount of spending solely towards recycled products, or "price preference" programs which provide a larger budget when recycled items are purchased. Additional regulations can target specific cases: in the United States, for example, the Environmental Protection Agency mandates the purchase of oil, paper tyres and building insulation from recycled or re-refined sources whenever possible.

The final government regulation towards increased demand is recycled product labeling. When producers are required to label their packaging with amount of recycled material in the product (including the packaging), consumers are better able to make educated choices. Consumers with sufficient buying power can then choose more environmentally conscious options, prompt producers to increase the amount of recycled material in their products, and indirectly increase demand. Standardized recycling labeling can also have a positive effect on supply of recyclates if the labeling includes information on how and where the product can be recycled.

Process

A number of different systems have been implemented to collect recyclates from the general waste stream. These systems tend to lie along the spectrum of trade-off between public convenience and government ease and expense. The three main categories of collection are "drop-off centres", "buy-back centres" and "curbside collection".

Drop-off centres require the waste producer to carry the recyclates to a central location, either an installed or mobile collection station or the reprocessing plant itself. They are the easiest type of collection to establish, but suffer from low and unpredictable throughput. **Buy-back centres** differ in that the cleaned recyclates are purchased, thus providing a clear incentive for use and creating a stable supply. The post-processed material can then be sold on, hopefully creating a profit. Unfortunately government subsidies are necessary to make buy-back centres a viable enterprise.

Curbside collection

Curbside collection encompasses many subtly different systems, which differ mostly on where in the process the recyclates are sorted and cleaned. The main categories are mixed waste collection, commingled recyclables and source separation. A waste collection vehicle generally picks up the waste.

At one end of the spectrum is mixed waste collection, in which all recyclates are collected and mixed in with the rest of the waste, and the desired material is then sorted out and cleaned at a central sorting facility. This results in a large amount of recyclable waste, paper especially, being too soiled to reprocess, but has advantages as well: the city need not pay for a separate collection of recyclates and no public education is needed. Any changes to which materials are recyclable is easy to accommodate as all sorting happens in a central location.

In a Commingled or Single-stream system, all recyclables for collection are mixed but kept separate from other waste. This greatly reduces the need for post-collection cleaning but does require public education on what materials are recyclable. Source separation is the other extreme, where each material is cleaned and sorted prior to collection. This method requires the least post-collection sorting and produces the purest recyclates, but incurs additional operating costs for collection of each separate material. An extensive public education program is also required, which must be successful if recyclate contamination is to be avoided. Source separation used to be the preferred method due to the high sorting costs incurred by commingled collection. Advances in sorting technology however, have lowered this overhead substantially—many areas which had developed source separation programs have since switched to commingled collection.

Sorting

Once commingled recyclates are collected and delivered to a central collection facility, the different types of materials must be sorted. This is done in a series of stages, many of which involve automated processes such that a truck-load of material can be fully sorted in less than an hour. Some plants can now sort the materials automatically, known as Single Stream. A 30% increase in recycling rates has been seen in the areas where these plants exist.

Initially, the commingled recyclates are removed from the collection vehicle and placed on a conveyor belt spread out in a single layer. Large pieces of cardboard and plastic bags are removed by hand at this stage, as they can cause later machinery to jam.

Next, automated machinery separates the recyclates by weight, splitting lighter paper and plastic from heavier glass and metal. Cardboard is removed from the mixed paper, and the most common types of plastic, PET (#1) and HDPE (#2) are collected. This separation is usually done by hand, but has become automated in some sorting centers: a spectroscopic scanner is used to differentiate between different types of paper and plastic based on the absorbed wavelengths, and subsequently divert each material into the proper collection channel.

Strong magnets are used to separate out ferrous metals such as iron, steel, and tin-plated steel cans ("tin cans"). Non-ferrous metals are ejected by magnetic eddy currents in which a rotating magnetic field induces an electric current around the aluminum cans, which in turn creates a magnetic eddy current inside the cans. This magnetic eddy current is repulsed by a large magnetic field, and the cans are ejected from the rest of the recyclate stream.

Energy Consumption and Saving

There is controversy on just how much energy is saved through recycling. The Energy Information Administration (EIA) states on its website that "a paper mill uses 40 percent less energy to make paper from recycled paper than it does to make paper from fresh lumber." Critics often argue that in the overall processes, it can take more energy to produce recycled products than it does to dispose them off in traditional landfill methods. This argument is followed from the curbside collection of recyclables, which critics note is often done by a second waste truck. Recycling proponents point out that a second timber or logging truck is eliminated when paper is collected for recycling.

It is difficult to determine the exact amount of energy consumed or produced in waste disposal processes. How much energy is used in recycling depends largely on the type of material being recycled and the process used to do so. Aluminum is generally agreed to use far less energy when recycled rather than being produced from scratch. The EPA states that "recycling aluminum cans, for example, saves 95 percent of the energy required to make the same amount of aluminum from its virgin source, bauxite."

Economist Steven Landsburg has suggested that the sole benefit of reducing landfill space is trumped by the energy needed and resulting pollution from the recycling process. Others, however, have calculated through life cycle assessment that producing recycled paper uses less energy and water than harvesting, pulping, processing, and transporting virgin trees. By using less recycled paper, additional energy is needed to create and maintain farmed forests until these forests are as self-sustainable as virgin forests.

Public policy analyst, James V. DeLong points out that recycling is a manufacturing process and many of the methods use more energy than they save. In addition to energy usage, he notes that recycling requires capital and labor while producing some waste. These processes need to be more efficient than production from original raw material and/or traditional garbage disposal in order for recycling to be the superior method.

Chapter 15

Solid Waste Management

The huge amount of municipal refuse generated daily demands for a highly efficient system to manage its collection and disposal. Our operations over the years have resulted in Singapore having one of the most efficiently operated refuse collection services in the region.

Public Waste Collection Scheme

The public waste collection scheme is built on a collective platform where pre-qualified waste collection companies competed to provide refuse collection services for the designated domestic and trade premises in the island's nine geographical sectors. Successful bidders are awarded tenders to serve the respective sectors for a period of 7 years. They are also required to provide door-to-door collection services for recyclable materials from the households in their sectors under the National Recycling Programme.

Licensed General Waste Collectors

The Agency also licenses and monitors the collection of refuse by licensed general waste collectors, which mainly serve commercial and industrial premises. It is an offence for any person or company to collect or transport waste as a business without a valid General Waste Collector Licence. Licensed general waste collectors can refer to 'The Code of Practice for Licensed Waste Collectors' for work protocols and guidelines on good practices in the waste collection business.

SOLID WASTE DISPOSAL SYSTEM

NEA's solid waste disposal infrastructure comprises four incineration plants - Ulu Pandan, Tuas, Senoko and Tuas South, and an offshore Semakau Landfill. Total of 2.63 million tonnes of waste was disposed off in 2008. About 2.45 million tonnes or 93.2% of the refuse was incinerated while the remaining refuse was landfilled.

Of the amount disposed off, 57% was generated from residential premises, food centres and markets. Commerical and industrial premises accounted for the remaining 43%. The average amount of domestic refuse generated per person per day in 2008 was 0.84 kilogrammes.

Enforcement on Illegal Dumping of Waste

Illegal dumping of waste happens mostly in secluded areas, under the cover of night. It poses serious potential risks to our environment that lead to pollution, public nuisances and public health hazards. It is also costly to clean up the dumped waste. NEA is determined to stamp out illegal dumping in Singapore.

SOLID WASTE DISPOSAL MANAGEMENT

Waste management is the collection, transport, processing, recycling or disposal, and monitoring of waste materials. The term usually relates to materials produced by human activity, and is generally undertaken to reduce their effect on health, the environment or aesthetics. Waste management is also carried out to recover resources from it. Waste management can involve solid, liquid, gaseous or radioactive substances, with different methods and fields of expertise for each.

Waste management practices differ for developed and developing nations, for urban and rural areas, and for residential and industrial producers. Management for non-hazardous residential and institutional waste in metropolitan areas is usually the responsibility of local government authorities, while management for non-hazardous commercial and industrial waste is usually the responsibility of the generator.

Disposal Methods

Landfill

Disposing of waste in a landfill involves burying the waste, and this remains a common practice in most countries. Landfills were often established in abandoned or unused quarries, mining voids or borrow pits. A properly-designed and well-managed landfill can be a hygienic and relatively inexpensive method of disposing of waste materials. Older, poorly-designed or poorly-managed landfills can create a number of adverse environmental impacts such as wind-blown litter, attraction of vermin, and generation of liquid leachate. Another common by-product of landfills is gas (mostly composed of methane and carbon dioxide) which is produced as organic waste, breaks down anaerobically. This gas can create odor problems, kill surface vegetation, and is a greenhouse gas.

Design characteristics of a modern landfill include methods to contain leachate such as clay or plastic lining material. Deposited waste is normally compacted to increase its density and stability, and covered to prevent attracting vermin (such as mice or rats). Many landfills also have landfill gas extraction systems installed to extract the landfill gas. Gas is pumped out of the landfill using perforated pipes and flared-off or burnt in a gas engine to generate electricity.

Incineration

Incineration is a disposal method that involves combustion of waste material. Incineration and other high temperature waste treatment systems are sometimes described as "thermal treatment". Incinerators convert waste materials into heat, gas, steam, and ash.

Incineration is carried out both on a small scale by individuals and on a large scale by industry. It is used to dispose off solid, liquid and gaseous waste. It is recognized as a practical method of disposing off certain hazardous waste materials (such as biological, medical waste). Incineration is a controversial method of waste disposal, due to issues such as emission of gaseous pollutants.

Incineration is common in countries such as Japan where land is more scarce, as these facilities generally do not require as much area as landfills. Waste-to-energy (WtE) or energy-from-waste (EfW) are broad terms for facilities that burn waste in a furnace or boiler to generate heat, steam and/or electricity. Combustion in an incinerator is not always perfect and there have been concerns about micro-pollutants in gaseous emissions from incinerator stacks. Particular concern has focused on some very persistent organics such as dioxins which may be created within the incinerator and which may have serious environmental consequences in the area immediately around the incinerator. On the other hand, this method produces heat that can be used as energy.

Recycling Methods

PVC, LDPE, PP, and PS are also recyclable, although these are not as commonly collected. These items are usually composed of a single type of material, making them relatively easy to recycle into new products. The recycling of complex products (such as computers and electronic equipment) is more difficult, due to the additional dismantling and separation required.

Biological Reprocessing

Waste materials that are organic in nature, such as plant material, food scraps, and paper products, can be recycled using biological composting and digestion processes to decompose the organic matter. The resulting organic material is then recycled as mulch or compost for agricultural or landscaping purposes. In addition, waste gas from the process (such as methane) can be captured and used for generating electricity. The intention of biological processing in waste management is to control and accelerate the natural process of decomposition of organic matter.

There are a large variety of composting and digestion methods and technologies varying in complexity from simple home compost heaps, to industrial-scale enclosed-vessel digestion of mixed domestic waste. Methods of biological decomposition are differentiated as being aerobic or anaerobic methods, though hybrids of the two methods also exist.

An example of waste management through composting is the Green Bin Program in Toronto, Canada, where household organic waste (such as kitchen scraps and plant cuttings) are collected in a dedicated container and then composted.

Energy Recovery

The energy content of waste products can be harnessed directly by using them as a direct combustion fuel, or indirectly by processing them into another type of fuel. Recycling through thermal treatment ranges from using waste as a fuel source for smoking or heating, to fuel for boilers to generate steam and electricity in a turbine. Pyrolysis and gasification are two related forms of thermal treatment where waste materials are heated to high temperatures with limited oxygen availability. The process typically occurs in a sealed vessel under high pressure. Pyrolysis of solid waste converts the material into solid, liquid and gas products. The liquid and gas can be burnt to produce energy or refined into other products. The solid residue (char) can be further refined into products such as activated carbon. Gasification and advanced Plasma arc gasification are used to convert organic materials directly into a synthetic gas (syngas) composed of carbon monoxide and hydrogen. The gas is then burnt to produce electricity and steam.

Avoidance and Reduction Methods

An important method of waste management is the prevention of waste material being created, also known as waste reduction. Methods of avoidance include reuse of second-hand products, repairing broken items instead of buying new, designing products to be refillable or reusable (such as cotton instead of plastic shopping bags), encouraging consumers to avoid using disposable products (such as disposable cutlery), removing any food/liquid remains from cans, packaging, and designing products that use less material to achieve the same purpose (for example, light-weighting of beverage cans).

Waste Handling and Transport

Waste collection methods vary widely between different countries and regions. Domestic waste collection services are often provided by local government authorities, or by private industry. Some areas, especially those in less developed countries, do not have a formal waste-collection system.

Technologies

Traditionally, the Waste Management Industry has been slow to adopt new technologies such as RFID tags, GPS and integrated software packages which enable better quality data to be collected without the use of estimation or manual data entry.

- Technologies like RFID tags are now being used to collect data on presentation rates for curb-side pick-ups which is useful when examining the usage of recycling bins or similar.
- Benefits of GPS tracking is particularly evident when considering the efficiency of *adhoc* pick-ups (like skip bins or dumpsters) where the collection is done on a consumer request basis.

- Integrated software packages are useful in aggregating this data for use in optimisation of operations for waste collection operations.

Waste Management Concepts

There are a number of concepts about waste management which vary in their usage between countries or regions. Some of the most general, widely-used concepts include:

The waste hierarchy refers to the "3 Rs" reduce, reuse and recycle, which classify waste management strategies according to their desirability in terms of waste minimization. The waste hierarchy remains the cornerstone of most waste minimization strategies. The aim of the waste hierarchy is to extract the maximum practical benefits from products and to generate the minimum amount of waste.

- Extended Producer Responsibility - Extended Producer Responsibility (EPR) is a strategy designed to promote the integration of all costs associated with products throughout their life cycle (including end-of-life disposal costs) into the market price of the product. Extended producer responsibility is meant to impose accountability over the entire life cycle of products and packaging introduced to the market. This means that firms which manufacture, import and/or sell products are required to be responsible for the products after their useful life as well as during manufacture.
- Polluter Pays Principle - The Polluter Pays Principle is a principle where the polluting party pays for the impact caused to the environment. With respect to waste management, this generally refers to the requirement for a waste generator to pay for appropriate disposal of the waste.

Education and Awareness

Education and awareness in the area of waste and waste management is increasingly important from a global perspective of resource management. The Talloires Declaration is a declaration for sustainability concerned about the unprecedented scale and speed of environmental pollution and degradation, and the depletion of natural resources. Local, regional, and global air pollution; accumulation and distribution of toxic wastes; destruction and depletion of forests, soil, and water; depletion of the ozone layer and emission of "green house" gases threaten the survival of humans and thousands of other living species, the integrity of the earth and its biodiversity, the security of nations, and the heritage of future generations. Several universities have implemented the Talloires Declaration by establishing environmental management and waste management programs, e.g., the waste management university project. University and vocational education are promoted by various organizations, e.g., WAMITAB and Chartered Institution of Wastes Management. Many supermarkets encourage customers to use their reverse vending machines to deposit used purchased containers and receive a refund from the recycling fees. Brands that manufacture such machines include Tomra and Envipco.

Sewage Treatment

Sewage treatment or domestic wastewater treatment is the process of removing contaminants from wastewater and household sewage, both runoff (effluents) and domestic. It includes physical, chemical and biological processes to remove physical, chemical and biological contaminants. Its objective is to produce a waste stream (or treated effluent) and a solid waste or sludge suitable for discharge or reuse back into the environment. This material is often inadvertently contaminated with many toxic organic and inorganic compounds.

Origin of Sewage

Sewage is created by residences, institutions, hospitals and commercial and industrial establishments. Raw influent (sewage) includes household waste liquid from toilets, baths, showers, kitchens, sinks, and so forth that is disposed off via sewers. In many areas, sewage also includes liquid waste from industry and commerce.

The separation and draining of household waste into grey water and black water is becoming more common in the developed world, with grey water being permitted to be used for watering plants or recycled for flushing toilets. A lot of sewage also includes some surface water from roofs or hard-standing areas. Municipal wastewater therefore includes residential, commercial and industrial liquid waste discharges and may include storm water runoff. Sewage systems capable of handling stormwater are known as combined systems or combined sewers. Such systems are usually avoided since they complicate and thereby reduce the efficiency of sewage treatment plants owing to their seasonality. The variability in flow also leads to often larger than necessary, and subsequently more expensive, treatment facilities. In addition, heavy storms that contribute more flows than the treatment plant can handle may overwhelm the sewage treatment system, causing a spill or overflow. It is preferable to have a separate storm drain system for storm water in areas that are developed with sewer systems.

As rainfall runs over the surface of roofs and the ground, it may pick up various contaminants including soil particles and other sediment, heavy metals, organic compounds, animal waste, and oil and grease. Some jurisdictions require storm water to receive some level of treatment before being discharged directly into waterways. Examples of treatment processes used for storm water include sedimentation basins, wetlands, buried concrete vaults with various kinds of filters, and vortex separators (to remove coarse solids).

Process Overview

Sewage can be treated close to where it is created (in septic tanks, biofilters or aerobic treatment systems), or collected and transported via a network of pipes and pump stations to a municipal treatment plant. Sewage collection and treatment is typically subject to local, state and federal regulations and standards. Industrial sources of wastewater often require specialized treatment processes.

Conventional sewage treatment involves three stages, called *primary*, *secondary* and *tertiary treatment*. First, the solids are separated from the wastewater stream. Then dissolved biological matter is progressively converted into a solid mass by using indigenous, water-borne micro organisms. Finally, the biological solids are neutralized, then disposed off or re-used, and the treated water may be disinfected chemically or physically (for example, by lagoons and microfiltration). The final effluent can be discharged into a stream, river, bay, lagoon or wetland, or it can be used for the irrigation of a golf course, green way or park. If it is sufficiently clean, it can also be used for groundwater recharge or agricultural purposes.

Process Flow Diagram for a Typical Large-Scale Treatment Plant

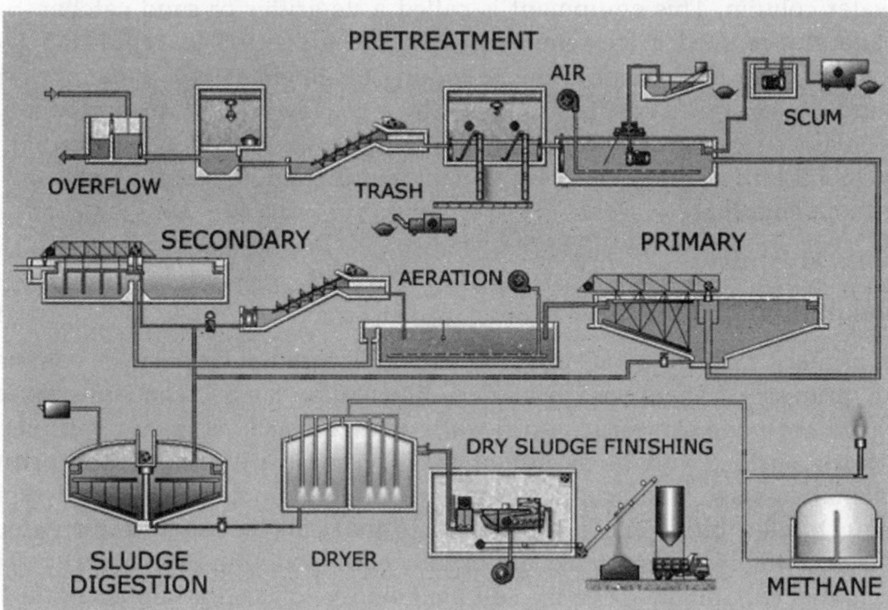

Pre-treatment

Pre-treatment removes the materials that can be easily collected from the raw wastewater and disposed off. The typical materials that are removed during pre-treatment include fats, oils, and greases (also referred to as FOG), sand, gravels and rocks (also referred to as grit), larger settleable solids and floating materials (such as rags and flushed feminine hygiene products). In developed countries, sophisticated equipment with remote operation and control are employed. The developing countries still rely on low cost equipment like manually cleaned screen etc.

Screening

The influent sewage water is strained to remove all large objects carried in the sewage stream such as rags, sticks, tampons, cans, fruits, etc. This is most commonly done with a manual or automated mechanically raked bar screen. The raking action of a mechanical bar screen is typically placed according to the accumulation on the bar screens and/or flow rate. The bar screen is used because large solids can damage or clog the equipment used later in the sewage treatment plant. The large solids can also hinder the biological process. The solids are collected and later disposed in a landfill or incineration.

Pre-treatment also typically includes a sand or grit channel or chamber where the velocity of the incoming wastewater is carefully controlled to allow sand grit and stones to settle, while keeping the majority of the suspended organic material in the water column. This equipment is called a de-gritter or sand catcher. Sand, grit, and stones need to be removed early in the process to avoid damage to pumps and other equipment in the remaining treatment stages. Sometimes there is a sand washer (grit classifier) followed by a conveyor that transports the sand to a container for disposal. The contents from the sand catcher may be fed into the incinerator in a sludge processing plant, but in many cases, the sand and grit is sent to a landfill.

Primary Treatment

Sedimentation

In the primary sedimentation stage, sewage flows through large tanks, commonly called "primary clarifiers" or "primary sedimentation tanks". The tanks are large enough that sludge can settle and floating material such as grease and oils can rise to the surface and be skimmed off. The main purpose of the primary sedimentation stage is to produce both a generally homogeneous liquid capable of being treated biologically and a sludge that can be separately treated or processed. Primary settling tanks are usually equipped with mechanically driven scrapers that continually drive the collected sludge towards a hopper in the base of the tank from where it can be pumped to further sludge treatment stages.

Secondary Treatment

Secondary treatment is designed to substantially degrade the biological content of the sewage such as are derived from human waste, food waste, soaps and detergent. The majority of municipal plants treat the settled sewage liquor using aerobic biological processes. For this to be effective, the biota requires both oxygen and a substrate on which to live. There are a number of ways in which this is done. In all of these methods, the bacteria and protozoa consume biodegradable soluble organic contaminants (e.g. sugars, fats, organic short-chain carbon molecules, etc.) and bind much of the less soluble fractions into floc. Secondary treatment systems are classified as:

- **fixed-film,** or
- suspended-growth.

Fixed-film treatment process including trickling filter and rotating biological contactors where the biomass grows on media and the sewage passes over its surface.

In **suspended-growth** systems, such as activated sludge, the biomass is well mixed with the sewage and can be operated in a smaller space than fixed-film systems that treat the same amount of water. However, fixed-film systems are more able to cope with drastic changes in the amount of biological material and can provide higher removal rates for organic material and suspended solids than suspended growth systems.

Roughing filters are intended to treat particularly strong or variable organic loads, typically industrial, to allow them to then be treated by conventional secondary treatment processes. Characteristics include typically tall, circular filters filled with open synthetic filter media to which wastewater is applied at a relatively high rate. They are designed to allow high hydraulic loading and a high flow-through of air. On larger installations, air is forced through the media using blowers. The resultant wastewater is usually within the normal range for conventional treatment processes.

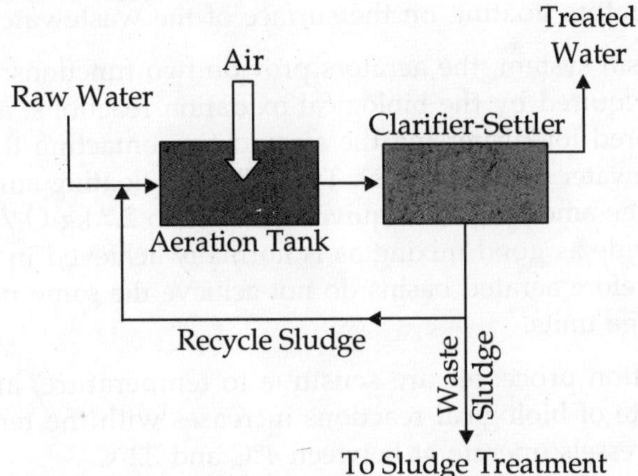

A generalized, schematic diagram of an activated sludge process.

Activated Sludge

In general, activated sludge plants encompass a variety of mechanisms and processes that use dissolved oxygen to promote the growth of biological floc that substantially removes organic material.

The process traps particulate material and can, under ideal conditions, convert ammonia to nitrite and nitrate and ultimately to nitrogen gas.

Surface-aerated basins

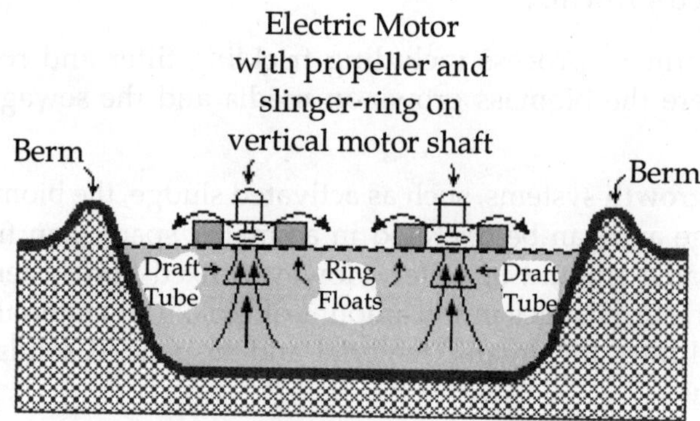

A Typical Surface-Aerated Basin (using motor-driven floating aerators)

Most biological oxidation processes for treating industrial wastewaters have in common the use of oxygen (or air) and microbial action. Surface-aerated basins achieve 80 to 90% removal of Biochemical Oxygen Demand with retention times of 1 to 10 days. The basins may range in depth from 1.5 to 5.0 metres and use motor-driven aerators floating on the surface of the wastewater.

In an aerated basin system, the aerators provide two functions: they transfer air into the basins required by the biological oxidation reactions, and they provide the mixing required for dispersing the air and for contacting the reactants (that is, oxygen, wastewater and microbes). Typically, the floating surface aerators are rated to deliver the amount of air equivalent to 1.8 to 2.7 kg O_2/kW·h. However, they do not provide as good mixing as is normally achieved in activated sludge systems and therefore aerated basins do not achieve the same performance level as activated sludge units.

Biological oxidation processes are sensitive to temperature, and between 0°C and 40°C, the rate of biological reactions increases with the temperature. Most surface aerated vessels operate at between 4°C and 32°C.

Filter Beds (Oxidizing Beds)

Trickling Filter

In older plants and plants receiving more variable loads, trickling filter beds are used where the settled sewage liquor is spread onto the surface of a deep bed made up of coke (carbonized coal), limestone chips or specially fabricated plastic media. Such media must have high surface areas to support the biofilms that form. The liquor is distributed through perforated rotating arms radiating from a central pivot. The distributed liquor trickles through this bed and is collected in drains at the base. These drains also provide a source of air which percolates up through the bed, keeping it aerobic. Biological films of bacteria, protozoa and fungi form on the media's surfaces and eat or otherwise reduce the organic content. This biofilm is grazed by insect larvae and worms which help to maintain an optimal thickness. Overloading of beds increases the thickness of the film leading to clogging of the filter media and ponding on the surface.

Biological Aerated Filters

Biological Aerated (or Anoxic) Filter (BAF) or Biofilters combine filtration with biological carbon reduction, nitrification or denitrification. BAF usually includes a reactor filled with a filter media. The media is either in suspension or supported by a gravel layer at the foot of the filter. The dual purpose of this media is to support highly active biomass that is attached to it and to filter suspended solids. Carbon reduction and ammonia conversion occurs in aerobic mode and sometimes achieved in a single reactor while nitrate conversion occurs in anoxic mode. BAF is operated either in upflow or downflow configuration depending on design specified by manufacturer.

Membrane Bioreactors

Membrane bioreactors (MBR) combines activated sludge treatment with a membrane liquid-solid separation process. The membrane component uses low pressure microfiltration or ultra filtration membranes and eliminates the need for clarification and tertiary filtration. The membranes are typically immersed in the aeration tank (however, some applications utilize a separate membrane tank). One of the key benefits of a membrane bioreactor system is that it effectively overcomes the limitations associated with poor settling of sludge in conventional activated sludge (CAS) processes. The technology permits bioreactor operation with considerably higher mixed liquor suspended solids (MLSS) concentration than CAS systems, which are limited by sludge settling. The process is typically operated at MLSS in the range of 8,000–12,000 mg/L, while CAS are operated in the range of 2,000–3,000 mg/L. The elevated biomass concentration in the membrane bioreactor process allows for very effective removal of both soluble and particulate biodegradable materials at higher loading rates. Thus increased Sludge Retention Times (SRTs)—usually exceeding 15 days—ensure complete nitrification even in extremely cold weather.

The cost of building and operating a MBR is usually higher than conventional wastewater treatment, however, as the technology has become increasingly popular and has gained wider acceptance throughout the industry, the life cycle costs have been steadily decreasing. The small footprint of MBR systems, and the high quality effluent produced, makes them particularly useful for water reuse applications.

Secondary Sedimentation

The final step in the secondary treatment stage is to settle out the biological floc or filter material and produce sewage water containing very low levels of organic material and suspended matter.

Rotating biological contactor

Rotating biological contactors (RBCs) are mechanical secondary treatment systems, which are robust and capable of withstanding surges in organic load. RBCs were first installed in Germany in 1960 and have since been developed and refined into a reliable operating unit. The rotating disks support the growth of bacteria and micro-organisms present in the sewage, which breakdown and stabilise organic pollutants. To be successful, micro organisms need both oxygen to live and food to grow. Oxygen is obtained from the atmosphere as the disks rotate. As the micro organisms grow, they build up on the media until they are sloughed off due to shear forces provided by the rotating discs in the sewage. Effluent from the RBC is then passed through final clarifiers where the micro organisms in suspension settle as a sludge. The sludge is withdrawn from the clarifier for further treatment.

A functionally similar biological filtering system has become popular as part of home aquarium filtration and purification. The aquarium water is drawn up out of the tank and then cascaded over a freely spinning corrugated fiber-mesh wheel before passing through a media filter and back into the aquarium. The spinning mesh wheel develops a biofilm coating of microorganisms that feed on the suspended wastes in the aquarium water and are also exposed to the atmosphere as the wheel rotates. This is especially good at removing waste urea and ammonia urinated into the aquarium water by the fish and other animals.

Tertiary Treatment

The purpose of tertiary treatment is to provide a final treatment stage to raise the effluent quality before it is discharged to the receiving environment (sea, river, lake, ground, etc.). More than one tertiary treatment process may be used at any treatment plant. If disinfection is practiced, it is always the final process. It is also called "effluent polishing".

Filtration

Sand filtration removes much of the residual suspended matter. Filtration over activated carbon removes residual toxins.

Lagooning

Lagooning provides settlement and further biological improvement through storage in large man-made ponds or lagoons. These lagoons are highly aerobic and colonization by native macrophytes, especially reeds, is often encouraged.

Small filter feeding invertebrates such as Daphnia and species of Rotifera greatly assist in the treatment by removing fine particulates.

Constructed Wetlands

Constructed wetlands include engineered reedbeds and a range of similar methodologies, all of which provide a high degree of aerobic biological improvement and can often be used instead of secondary treatment for small communities, (also see phytoremediation). One example is a small reedbed used to clean the drainage from the elephants' enclosure at Chester Zoo in England.

Nutrient Removal

Wastewater may contain high levels of the nutrients nitrogen and phosphorus. Excessive release to the environment can lead to a build-up of nutrients called eutrophication, which can in turn, encourage the overgrowth of weeds, algae, and cyanobacteria (blue-green algae). This may cause an algal bloom, a rapid growth in the population of algae. The algae numbers are unsustainable and eventually most of them die. The decomposition of the algae by bacteria uses up so much of oxygen in the water that most or all of the animals die, which creates more organic matter for the bacteria to decompose. In addition to causing deoxygenation, some algal species produce toxins that contaminate drinking water supplies. Different treatment processes are required to remove nitrogen and phosphorus.

Nitrogen Removal

The removal of nitrogen is effected through the biological oxidation of nitrogen from ammonia (nitrification) to nitrate, followed by denitrification, the reduction of nitrate to nitrogen gas. Nitrogen gas is released to the atmosphere and thus removed from the water.

Nitrification itself is a two-step aerobic process, each step facilitated by a different type of bacteria. The oxidation of ammonia (NH_3) to nitrite (NO_2^-) is most often facilitated by *Nitrosomonas* spp. (nitroso referring to the formation of a nitroso functional group). Nitrite oxidation to nitrate (NO_3^-), though traditionally believed to be facilitated by *Nitrobacter* spp. (nitro referring the formation of a nitro functional group), is now known to be facilitated in the environment almost exclusively by *Nitrospira* spp.

Denitrification requires anoxic conditions to encourage the appropriate biological communities to form. It is facilitated by a wide diversity of bacteria. Sand filters, lagooning and reed beds can all be used to reduce nitrogen, but the activated sludge process (if designed well) can do the job most easily. Since denitrification is the reduction of nitrate to dinitrogen gas, an electron donor is needed. This may, depend on the wastewater, organic matter (from faeces), sulfide, or an added donor like methanol.

Sometimes the conversion of toxic ammonia to nitrate alone is referred to as tertiary treatment.

Phosphorus Removal

Phosphorus removal is important as it is a limiting nutrient for algae growth in many fresh water systems. It is also particularly important for water reuse systems where high phosphorus concentrations may lead to fouling of downstream equipment such as reverse osmosis.

Phosphorus can be removed biologically in a process called enhanced biological phosphorus removal. In this process, specific bacteria, called polyphosphate accumulating organisms (PAOs), are selectively enriched and accumulate large quantities of phosphorus within their cells (up to 20% of their mass). When the biomass enriched in these bacteria is separated from the treated water, these biosolids have a high fertilizer value.

Phosphorus removal can also be achieved by chemical precipitation, usually with salts of iron (e.g., ferric chloride), aluminium (e.g., alum), or lime. This may lead to excessive sludge production as hydroxides precipitates and the added chemicals can be expensive. Despite this, chemical phosphorus removal requires significantly smaller equipment footprint than biological removal, is easier to operate and is often more reliable than biological phosphorus removal.

Once removed, phosphorus, in the form of a phosphate rich sludge, may be landfilled or, if in suitable condition, resold for use in fertilizer.

Disinfection

The purpose of disinfection in the treatment of wastewater is to substantially reduce the number of microorganisms in the water to be discharged back into the environment. The effectiveness of disinfection depends on the quality of the water being treated (e.g., cloudiness, pH, etc.), the type of disinfection being used, the disinfectant dosage (concentration and time), and other environmental variables. Cloudy water will be treated less successfully since solid matter can shield organisms, especially from ultraviolet light or if contact times are low. Generally, short contact times, low doses and high flows all militate against effective disinfection. Common methods of disinfection include ozone, chlorine, or ultraviolet light. Chloramine, which is used for drinking water, is not used in wastewater treatment because of its persistence.

Chlorination remains the most common form of wastewater disinfection in North America due to its low cost and long-term history of effectiveness. One disadvantage is that chlorination of residual organic material can generate chlorinated-organic compounds that may be carcinogenic or harmful to the environment. Residual chlorine or chloramines may also be capable of chlorinating organic material in the natural aquatic environment. Further, because residual chlorine is toxic to aquatic species, the treated effluent must also be chemically dechlorinated, adding to the complexity and cost of treatment.

Ultraviolet (UV) light can be used instead of chlorine, iodine, or other chemicals. Because no chemicals are used, the treated water has no adverse effect on organisms that later consume it, as may be the case with other methods. UV radiation causes damage to the genetic structure of bacteria, viruses, and other pathogens, making them incapable of reproduction. The key disadvantages of UV disinfection are the need for frequent lamp maintenance and replacement and the need for a highly treated effluent to ensure that the target microorganisms are not shielded from the UV radiation (i.e., any solids present in the treated effluent may protect microorganisms from the UV light). In the United Kingdom, light is becoming the most common means of disinfection because of the concerns about the impacts of chlorine in chlorinating residual organics in the wastewater and in chlorinating organics in the receiving water. Edmonton and Calgary, Alberta, Canada also use UV light for their effluent water disinfection.

Ozone (O_3) is generated by passing oxygen (O_2) through a high voltage potential resulting in a third oxygen atom becoming attached and forming O_3. Ozone is very unstable and reactive and oxidizes most organic material it comes in contact with, thereby destroying many pathogenic microorganisms. Ozone is considered to be safer than chlorine because, unlike chlorine which has to be stored on site (highly poisonous in the event of an accidental release), ozone is generated onsite as needed. Ozonation also produces fewer disinfection by-products than chlorination. A disadvantage of ozone disinfection is the high cost of the ozone generation equipment and the requirements for special operators.

Package Plants and Batch Reactors

In order to use less space, treat difficult waste, deal with intermittent flow or achieve higher environmental standards, a number of designs of hybrid treatment plants have been produced. Such plants often combine all or at least two stages of the three main treatment stages into one combined stage. In the UK, where a large number of sewage treatment plants serve small populations, package plants are a viable alternative to building discrete structures for each process stage.

One type of system that combines secondary treatment and settlement is the sequencing batch reactor (SBR). Typically, activated sludge is mixed with raw incoming sewage and mixed and aerated. The resultant mixture is then allowed to settle producing a high quality effluent. The settled sludge is run off and re-aerated before a proportion is returned to the head of the works. SBR plants are now being deployed in many parts of the world including North Liberty, Iowa, and Llanasa, North Wales.

The disadvantage of such processes is that precise control of timing, mixing and aeration is required. This precision is usually achieved by computer controls linked to many sensors in the plant. Such a complex, fragile system is unsuited to places where such controls may be unreliable, or poorly maintained, or where the power supply may be intermittent.

Package plants may be referred to as *high charged* or *low charged*. This refers to the way the biological load is processed. In *high charged* systems, the biological stage is presented with a high organic load and the combined floc and organic material is then oxygenated for a few hours before being charged again with a new load. In the *low charged* system the biological stage contains a low organic load and is combined with flocculate for a relatively long time.

Electricity Use

Wastewater treatment plants are, along with water treatment plants, often the largest users of energy in a community. These processes account for 3% of electricity use in the United States. As an example, Nazareth Wastewater Treatment Plant in Nazareth, PA uses 1.6 million kWh per year to process 365 million gallons of wastewater. For each gallon of water treated, the electric power used is equivalent to leaving a 100 W equivalent (23W) Compact Fluorescent Lightbulb on for 11.4 minutes, or a 60 W equivalent (14W) bulb for 18.8 minutes.

Sludge Treatment and Disposal

The sludges accumulated in a wastewater treatment process must be treated and disposed off in a safe and effective manner. The purpose of digestion is to reduce the amount of organic matter and the number of disease-causing microorganisms present in the solids. The most common treatment options include anaerobic digestion, aerobic digestion, and composting.

Choice of a wastewater solid treatment method depends on the amount of solids generated and other site-specific conditions. However, in general, composting is most often applied to smaller-scale applications followed by aerobic digestion and then lastly anaerobic digestion for the larger-scale municipal applications.

Anaerobic Digestion

Anaerobic digestion is a bacterial process that is carried out in the absence of oxygen. The process can either be *thermophilic* digestion, in which sludge is fermented in tanks at a temperature of 55°C, or *mesophilic*, at a temperature of around 36°C. Though allowing shorter retention time (and thus smaller tanks), thermophilic digestion is more expensive in terms of energy consumption for heating the sludge.

One major feature of anaerobic digestion is the production of biogas, which can be used in generators for electricity production and/or in boilers for heating purposes.

Aerobic Digestion

Aerobic digestion is a bacterial process occurring in the presence of oxygen. Under aerobic conditions, bacteria rapidly consume organic matter and convert it into carbon dioxide. The operating costs are characteristically much greater for aerobic digestion because of the energy costs needed to add oxygen to the process.

Composting

Composting is also an aerobic process that involves mixing the sludge with sources of carbon such as sawdust, straw or wood chips. In the presence of oxygen, bacteria digest both the wastewater solids and the added carbon source and, in doing so, produce a large amount of heat.

Sludge Disposal

When a liquid sludge is produced, further treatment may be required to make it suitable for final disposal. Typically, sludges are thickened (dewatered) to reduce the volumes transported off-site for disposal. There is no process which completely eliminates the need to dispose off biosolids. There is, however, an additional step some cities are taking to superheat the wastewater sludge and convert it into small pelletized granules that are high in nitrogen and other organic materials. In New York City, for example, several sewage treatment plants have dewatering facilities that use large centrifuges along with the addition of chemicals such as polymer to further remove liquid from the sludge. The removed fluid, called centrate, is typically reintroduced into the wastewater process. The product which is left is called "cake" and that is picked up by companies which turn it into fertilizer pellets. This product is then sold to local farmers and turf farms as a soil amendment or fertilizer, reducing the amount of space required to dispose off sludge in landfills.

Treatment in the Receiving Environment

Many processes in a wastewater treatment plant are designed to mimic the natural treatment processes that occur in the environment, whether that environment is a natural water body or the ground. If not overloaded, bacteria in the environment will consume organic contaminants, although this will reduce the levels of oxygen in the water and may significantly change the overall ecology of the receiving water. Native bacterial populations feed on the organic contaminants, and the numbers of disease-causing microorganisms are reduced by natural environmental conditions such as predation or exposure to ultraviolet radiation. Consequently, in cases where the receiving environment provides a high level of dilution, a high degree of wastewater treatment may not be required. However, recent evidence has demonstrated that very low levels of certain contaminants in wastewater, including hormones (from animal husbandry and residue from human hormonal contraception methods) and synthetic materials such as phthalates that mimic hormones in their action, can have an unpredictable adverse impact on the natural biota and potentially on humans if the water is re-used for drinking water. In the US and EU, uncontrolled discharges of wastewater to the environment are not permitted under law, and strict water quality requirements are to be met. A significant threat in the coming decades will be the increasing uncontrolled discharges of wastewater within rapidly developing countries.

Sewage Treatment in Developing Countries

There are few reliable figures on the share of the wastewater collected in sewers that is being treated in the world. In many developing countries, the bulk of domestic and industrial wastewater is discharged without any treatment or after primary treatment only. In Latin America, about 15% of collected wastewater passes through treatment plants (with varying levels of actual treatment). In Venezuela, a below average country in South America with respect to wastewater treatment, 97 percent of the country's sewage is discharged raw into the environment.

In a relatively developed Middle Eastern country such as Iran, Tehran's majority of population has totally untreated sewage injected to the city's groundwater. Israel has also aggressively pursued the use of treated sewer water for irrigation. In 2008, agriculture in Israel consumed 500 million cubic metres of potable water and an equal amount of treated sewer water. The country plans to provide a further 200 million cubic metres of recycled sewer water and build more desalination plants to supply even more water.

Most of sub-Saharan Africa is without wastewater treatment. Water utilities in developing countries are chronically underfunded because of low water tariffs, the inexistence of sanitation tariffs in many cases, low billing efficiency (i.e., many users that are billed do not pay) and poor operational efficiency (i.e., there are overly high levels of staff, there are high physical losses, and many users have illegal connections and are thus not being billed). In addition, wastewater treatment typically is the process within the utility that receives the least attention, partly because enforcement of environmental standards is poor. As a result of all these factors, operation and maintenance of many wastewater treatment plants is poor. This is evidenced by the frequent breakdown of equipment, shutdown of electrically operated equipment due to power outages or to reduce costs, and sedimentation due to lack of sludge removal.

Developing countries as diverse as Egypt, Algeria, China or Colombia have invested substantial sums in wastewater treatment without achieving a significant impact in terms of environmental improvement. Even if wastewater treatment plants are properly operating, it can be argued that the environmental impact is limited in cases where the assimilative capacity of the receiving waters (ocean with strong currents or large rivers) is high, as it is often the case.

Bioindicator

Caddisfly (order *Trichoptera*), a macroinvertebrate is used as an indicator of water quality.

Bioindicators are species used to monitor the health of an environment or ecosystem. They are any biological species or group of species whose function, population, or status can be used to determine ecosystem or environmental integrity. An example of such a group are the copepods and other small water crustaceans present in many water bodies. Such organisms are monitored for changes (biochemical, physiological, or behavioural) that may indicate a problem within their ecosystem. Bioindicators can tell us about the cumulative effects of different pollutants in the ecosystem and about how long a problem may have been present, which physical and chemical testing cannot.

Depending on the organism selected and their use, there are several types of bioindicators which are given below.

Plant Indicators

The presence or absence of certain plant or other vegetative life in an ecosystem can provide important clues about the health of the environment: environmental preservation.

- Lichens are organisms comprising both fungi and algae. (They are not plants.) Lichens are found on rocks and tree trunks, and they respond to environmental changes in forests, including changes in forest structure—conservation biology, air quality, and climate. The disappearance of lichens in a forest may indicate environmental stresses, such as high levels of sulfur dioxide, sulfur-based pollutants, and nitrogen oxides.
- The composition and total biomass of algal species in aquatic systems serves as an important metric for organic pollution and nutrient loading such as nitrogen and phosphorus.

Animal Indicators and Toxins

An increase or decrease in an animal population may indicate damage to the ecosystem caused by pollution. For example, if pollution causes the depletion of important food sources, animal species dependent upon these food sources will also be reduced in number: population decline. Overpopulation, can be the result of opportunistic species growth. In addition to monitoring the size and number of certain species, other mechanisms of animal indication include monitoring the concentration of toxins in animal tissues, or monitoring the rate at which deformities arise in animal populations.

Microbial Indicators and Chemical Pollutants

Microorganisms can be used as indicators of aquatic or terrestrial ecosystem health. Found in large quantities, microorganisms are easier to sample than other organisms. Some microorganisms will produce new proteins, called stress proteins, when exposed to contaminants like cadmium. These stress proteins can be used as an early warning system to detect high levels of pollution.

Macroinvertebrate Bioindicators

Macroinvertebrates are useful and convenient indicators of the ecological health of a water body or river. They are almost always present, and are easy to sample and identify. The sensitivity of the range of macroinvertebrates found will enable an objective judgement of the ecological condition to be made.

In Australia, the SIGNAL method has been developed and is used by researchers and community Waterwatch groups to monitor water health.

In the United States, the Environmental Protection Agency (EPA) has published *Rapid Bioassessment Protocols,* based on macroinvertebrates, as well as periphyton and fish. These protocols are used by many federal, state and local government agencies to design biosurveys for assessment of water quality. The species identification procedures are conducted in the field without the use of specialized equipment, and the techniques can be easily taught in volunteer training sessions.

Chapter 16

Agencies Related to Protection of Environmental Health

WORLD HEALTH ORGANIZATION

The **World Health Organization (WHO)** is a specialized agency of the United Nations (UN) that acts as a co-ordinating authority on international public health. Established on 7 April 1948, and head-quartered in Geneva, Switzerland, the agency inherited the mandate and resources of its predecessor, the Health Organization, which had been an agency of the League of Nations.

Constitution and History

The WHO's constitution states that its objective "is the attainment by all peoples of the highest possible level of health." Its major task is to combat disease, especially key infectious diseases, and to promote the general health of the people of the world.

The World Health Organization (WHO) is one of the original agencies of the United Nations, its constitution formally coming into force on 7 April 1948, when it was ratified by the 26th member state. Prior to this, its operations as well as the remaining activities of the League of Nations Health Organization, were under the control of an Interim Commission following an International Health Conference in the summet of 1946. The transfer was authorized by a Resolution of the General Assembly. The epidemiological service of the French *Office International d'Hygiène Publique* was incorporated into the Interim Commission of the World Health Organization on 1 January 1947.

Activities

As well as co-ordinating international efforts to monitor outbreaks of infectious diseases, such as SARS, malaria, swine flu, and AIDS, the WHO also sponsors programs to prevent and treat such diseases. The WHO supports the development and distribution of safe and effective vaccines, pharmaceutical diagnostics and drugs. After over 2 decades of fighting smallpox, the WHO declared in 1980 that the disease had been eradicated - the first disease in history to be eliminated by the human efforts.

The WHO aims to eradicate polio within the next few years. The organization has already endorsed the world's first official HIV/AIDS Toolkit for Zimbabwe (from 3 October 2006), making it an international standard.

In addition to its work in eradicating disease, the WHO also carries out various health-related campaigns – for example, to boost the consumption of fruits and vegetables worldwide and to discourage tobacco use.

Experts met at the WHO head-quarters in Geneva in February, 2007, and reported that their work on pandemic influenza vaccine development had achieved encouraging progress. More than 40 clinical trials have been completed or are ongoing. Most have focused on healthy adults. Some companies, after completing safety analyses in adults, have initiated clinical trials in the elderly and in children. All vaccines so far appear to be safe and well-tolerated in all age groups tested.

The WHO also conducts research; for example, whether the electromagnetic field surrounding cell phones has an impact on health. Some of this work can be controversial, as illustrated by the April, 2003, joint WHO/FAO report, which recommended that sugar should form no more than 10% of a healthy diet. This report led to lobbying by the sugar industry against the recommendation, to which the WHO/FAO responded by including in the report the statement "The Consultation recognized that a population goal for free sugars of less than 10% of total energy is controversial", but also stood by its recommendation based upon its own analysis of scientific studies.

Publishing

- International Classification of Diseases (ICD) is a widely followed publication. The tenth revision of the ICD, also known as ICD-10, was released in 1992 and a searchable version is available online on the WHO website. The WHO does not permit simultaneous classification in two separate areas.
- The Annual World Health Report, first published in 1995, is the WHO's leading publication. Each year the report combines an expert assessment of global health, including statistics relating to all countries, with a focus on a specific subject. The World Health Report 2007 - A safer future: global public health security in the 21st century was published on August 23, 2006.
- A model list of essential medicines that all countries' health-care systems should make available and affordable to the general population.
- Global plan of action on workers' health is a draft to protect and promote health in the workplace, to improve the performance of and access to occupational health services, and to incorporate workers' health into other policies. The WHO has emphasized the effort because, despite the availability of effective interventions to prevent occupational hazards, large gaps exist between and within countries with regard to the health status of workers and their exposure to occupational risks. According to the WHO, only a small minority of the global workforce has access to occupational health services. The action plan deals with aspects of workers' health, including primary prevention of occupational hazards, protection and promotion of health at work, employment conditions, and a better response from health

systems to workers' health.

- Health Sciences Online is a non-profit online health information resource from the World Health Organization.

The WHO website: A guide to statistical information at WHO and has an online version of the most recent WHO health statistics.

According to the WHO Programme on Health Statistics:

The production and dissemination of health statistics for health action at country, regional and global levels is a core WHO activity mandated to WHO by its Member States in its Constitution. WHO produced figures carry great weight in national and international resource allocation, policy making and programming, based on its reputation as "unbiased" (impartial and fair), global (not belonging to any camp), and technically competent (consulting leading research and policy institutions and individuals).

Programs

- The Yellow Card or Carte Jaune, is an international certificate of vaccination (ICV) issued by the WHO. It is recognised internationally and may be required for entry to certain countries where there are increased health risks for travellers. The Yellow Card should be kept in the holder's passport as it is a medical passport of sorts.
- International Radon Project
- Global Initiative for Emergency and Essential Surgical Care
- World Health Day is celebrated every year on 7 April.

Conventions

- Single Convention on Narcotic Drugs
- Convention on Psychotropic Substances
- World Health Organization Framework Convention on Tobacco Control

Private Sector Partnerships

- Aeras: Aeras Global TB Vaccine Foundation
- DNDi: Drugs for Neglected Diseases Initiative
- FIND: Foundation for Innovative New Diagnostics
- IAVI: International AIDS Vaccine Initiative
- IOWH: Institute for One World Health
- IPM: International Partnership for Microbicides
- MMV: Medicines for Malaria Venture
- MVI: Malaria Vaccine Initiative
- PATH: Program for Appropriate Technology in Health

- PDVI: Pediatric Dengue Vaccine Initiative
- TB Alliance: Global Alliance for TB Drug Development

Structure

The WHO has 193 Member States, including all UN Member States except Liechtenstein, and 2 non-UN members, Niue and the Cook Islands. Territories that are not UN Member States may join as Associate Members (with full information but limited participation and voting rights) if approved by an Assembly vote: Puerto Rico and Tokelau are Associate Members. Entities may also be granted observer status: examples include the Palestine Liberation Organization, Vatican City, and Chinese Taipei.

WHO Member States appoint delegations to the World Health Assembly, WHO's supreme decision-making body. All UN member states are eligible for WHO membership, and according to the WHO website, "Other countries may be admitted as members when their application has been approved by a simple majority vote of the World Health Assembly."

The WHO Assembly generally meets in May each year. In addition to appointing the Director General every five years, the Assembly considers the financial policies of the Organization and reviews and approves the proposed programme budget. The Assembly elects 34 members, technically qualified in the field of health, to the Executive Board for three-year terms. The main functions of the Board are to carry out the decisions and policies of the Assembly, to advise it and to facilitate its work in general.

The WHO is financed by contributions from member states and from donors. In the recent years, the WHO's work has involved more collaboration; there are currently around 80 such partnerships with NGOs and the pharmaceutical industry, as well as with foundations such as the Bill and Melinda Gates Foundation and the Rockefeller Foundation. Voluntary contributions to the WHO from national and local governments, foundations and NGOs, other UN organizations, and the private sector, now exceed that of assessed contributions (dues) from the 193 member nations.

Regional Offices

Regional offices and regions of the WHO: Africa; HQ: Brazzaville, Congo Americas; HQ: Washington, D.C., USA Eastern Med.; HQ: Cairo, Egypt Europe; HQ: Copenhagen, Denmark South East Asia; HQ: New Delhi, India Western Pacific; HQ: Manila, Philippines.

Uncharacteristically for a UN Agency, the six Regional Offices of the WHO enjoy remarkable autonomy. Each Regional Office is headed by a Regional Director (RD), who is elected by the Regional Committee for a once-renewable five-year term. The name of the RD-elect is transmitted to the WHO Executive Board in

Geneva, which proceeds to confirm the appointment. It is rare that an elected Regional Director is not confirmed.

Each Regional Committee of the WHO consists of all the Health Department heads, in all the governments of the countries that constitute the Region. Aside from electing the Regional Director, the Regional Committee is also in charge of setting the guidelines for the implementation, within the region of the Health and other policies adopted by the World Health Assembly. The Regional Committee also serves as a progress review board for the actions of the WHO within the Region.

The Regional Director is effectively the head of the WHO for his or her Region. The RD manages and/or supervises a staff of health and other experts at the regional head quarters and in specialized centers. The RD is also the direct supervising authority – concomitantly with the WHO Director General – of all the heads of WHO country offices, known as WHO Representatives, within the Region.

The Regional Offices are:

- Regional Office for Africa (AFRO), with heaa quarters in Brazzaville, Republic of Congo. AFRO includes most of Africa, with the exception of Egypt, Sudan, Tunisia, the Libyan Arab Jamahiriya, and Morocco which belong to EMRO. Somalia is also not counted as it does not have an official government, though it is in the process of getting one.
- Regional Office for Europe (EURO), with head-quarters in Copenhagen, Denmark.
- Regional Office for South East Asia (SEARO), with head-quarters in New Delhi, India. North Korea is served by SEARO.
- Regional Office for the Eastern Mediterranean (EMRO), with head-quarters in Cairo, Egypt. EMRO includes the countries of Africa, and particularly in the Maghreb, that are not included in AFRO, as well as the countries of the Middle East, except for Israel.
- Regional Office for Western Pacific (WPRO), with head-quarters in Manila, Philippines. WPRO covers all the Asian countries not served by SEARO and EMRO, and all the countries in Oceania. South Korea is served by WPRO.
- Regional Office for the Americas (AMRO), with headquarters in Washington, D.C., USA. It is better known as the Pan American Health Organization (PAHO). Since it predates the establishment of WHO, PAHO is by far the most autonomous of the 6 regional offices.

Country Offices

The World Health Organization operates 147 country and liaison offices in all its regions. The presence of a country office is generally motivated by a need stated by the member country. There will generally be one WHO country office in the capital, occasionally accompanied by satellite-offices in the provinces or sub-regions of the country in question.

The country office is headed by a WHO Representative (WR), who is a trained physician, not a national of that country, who holds diplomatic rank and has due privileges and immunities similar to those of an Ambassador Extraordinary and Plenipotentiary. In most countries, the WR (like Representatives of other UN agencies) is *de facto* and/or *de jure* treated like an Ambassador - the distinction here being that instead of being an Ambassador of one sovereign country to another, the WR is a senior UN civil servant, who serves as the "Ambassador" of the WHO to the country to which he or she is accredited. Hence, the title of *Resident Representative,* or simply *Representative.*

The country office consists of the WR, and several health and other experts, both foreign and local, as well as the necessary support staff. The main functions of WHO country offices include being the primary advisor of that country's government in matters of health and pharmaceutical policies.

International liaison offices serve largely the same purpose as country offices, but generally on a smaller scale. These are often found in countries that want WHO presence and co-operation, but do not have the major health system flaws that require the presence of a full-blown country office. Liaison offices are headed by a liaison officer, who is a national from that particular country, without diplomatic immunity.

People

Nancy Brinker was appointed Goodwill Ambassador for Cancer Control by Director General, Margaret Chan on May 26th.

Staffing

The World Health Organization is an agency of the United Nations and as such shares a core of common personnel policy with other agencies.

The World Health Organization has recently banned the recruitment of cigarette smokers, to promote the principle of a tobacco-free work environment. The World Health Organization(WHO) successfully rallied 168 countries to sign the Framework Convention on Tobacco Control in 2003. The Convention is designed to push for effective legislation and its enforcement in all countries to reduce the harmful effects of tobacco. On August 28, 2005, the National People's Congress of China signed the Convention.

United Nations (UN)	
UN System	Member states ·Observers ·General Assembly ·Security Council ·Economic and Social Council ·Secretariat (Secretary-General) ·Trusteeship Council ·International Court of Justice
Major offices	Headquarters (New York) ·Geneva ·Nairobi ·Vienna
Programmes and agencies	FAO ·ICAO ·ILO ·IMO ·IPCC ·IAEA ·UNIDO ·ITU ·UNAIDS ·SCSL ·UNCTAD ·UNCITRAL ·UNDCP ·UNDP ·UNEP ·UNESCO ·UNODC ·UNFIP ·UNIFEM ·UNFPA ·OHCHR ·UNHCR ·UNHRC ·UN-HABITAT ·UNICEF ·UNITAR ·UNOSAT ·UNRWA ·UNWTO ·UPU ·WFP ·WHO ·WMO
Resolutions	General Assembly (UDHR) ·Security Council
Related topics	Bretton Woods system ·Flag ·History ·ICC ·League of Nations ·OPCW ·Peacekeeping missions ·Charter ·Global Compact ·Treaty Series ·UN Day

Environmental Protection Agency (EPA)

The U.S. Environmental Protection Agency (EPA or sometimes USEPA) is an agency of the federal government of the United States charged to regulate chemicals and protect human health by safeguarding the natural environment: air, water, and land. The EPA was proposed by President Richard, Nixon and began its operations on December 2, 1970, when its establishment was passed by Congress, and signed into law by President Nixon, and has since been chiefly responsible for the environmental policy of the United States. It is led by its Administrator, who is appointed by the President of the United States. The EPA is not a Cabinet agency, but the Administrator is normally given cabinet rank. Lisa P. Jackson is the current Administrator. The agency has approximately 18,000 full-time employees.

The EPA employs 17,000 people in headquarters program offices, 10 regional offices, and 27 laboratories across the country. More than half of its staff are engineers, scientists, and environmental protection specialists; other groups include legal, public affairs, financial, and computer specialists.

The agency conducts environmental assessment, research, and education. It has the primary responsibility for setting and enforcing national standards under a variety of environmental laws, in consultation with state, tribal, and local governments. It delegates some permitting, monitoring, and enforcement responsibility to U.S. States and Native American tribes. EPA enforcement powers include fines, sanctions and other measures.

The agency also works with industries and all levels of government in a wide variety of voluntary pollution prevention programs and energy conservation efforts.

History

On July 9, 1970, President Nixon transmitted Reorganization Plan No. 3 to the United States Congress by executive order, creating the EPA as a single, independent agency from a number of smaller arms of different federal agencies. Prior to the establishment of the EPA, the federal government was not structured to comprehensively regulate the pollutants which harm human health and degrade the environment. The EPA was assigned the task of repairing the damage already done to the natural environment and to establish new criteria to guide Americans in making a cleaner, safer America.

Related Legislation

The legislation here is general environmental protection legislation, and may also apply to other units of the government, including the Department of the Interior and the Department of Agriculture.

Air

- 1955 - Air Pollution Control Act PL 84-159
- 1963 - Clean Air Act PL 88-206
- 1965 - Motor Vehicle Air Pollution Control Act PL 89-272
- 1966 - Clean Air Act Amendments PL 89-675
- 1967 - Air Quality Act PL 90-148
- 1969 - National Environmental Policy Act PL 91-190
- 1970 - Clean Air Act Extension PL 91-604
- 1976 - Toxic Substances Control Act PL 94-469
- 1977 - Clean Air Act Amendments PL 95-95
- 1990 - Clean Air Act Amendments PL 101-549

Water

- 1948 - Water Pollution Control Act PL 80-845
- 1965 - Water Quality Act PL 89-234
- 1966 - Clean Waters Restoration Act PL 89-753
- 1969 - National Environmental Policy Act PL 91-190
- 1970 - Water Quality Improvement Act PL 91-224
- 1972 - Federal Water Pollution Control Amendments of 1972 PL 92-500
- 1974 - Safe Drinking Water Act PL 93-523
- 1976 - Toxic Substances Control Act PL 94-469
- 1977 - Clean Water Act PL 95-217
- 1987 - Water Quality Act PL 100-4
- 1996 - Safe Drinking Water Act Amendments of 1996 PL 104-182

Land

- 1947 - Federal Insecticide, Fungicide, and Rodenticide Act
- 1964 - Wilderness Act PL 88-577
- 1968 - Scenic Rivers Preservation Act PL 90-542
- 1969 - National Environmental Policy Act PL 91-190
- 1970 - Wilderness Act PL 91-504
- 1977 - Surface Mining Control and Reclamation Act PL 95-87
- 1978 - Wilderness Act PL 98-625
- 1980 - Alaska Land Protection Act PL 96-487
- 1994 - California Desert Protection Act PL 103-433

Endangered species

- 1946 - Co-ordination Act PL 79-732
- 1966 - Endangered Species Preservation Act PL 89-669
- 1969 - Endangered Species Conservation Act PL 91-135
- 1972 - Marine Mammal Protection Act PL 92-522
- 1973 - Endangered Species Act PL 93-205

Hazardous waste

- 1965 - Solid Waste Disposal Act PL 89-272
- 1969 - National Environmental Policy Act PL 91-190
- 1970 - Resource Recovery Act PL 91-512
- 1976 - Resource Conservation and Recovery Act PL 94-580
- 1980 - Comprehensive Environmental Response, Compensation, and Liability Act ("Superfund") PL 96-510
- 1982 - Nuclear Waste Repository Act PL 97-425
- 1984 - Hazardous and Solid Wastes Amendments Act PL 98-616
- 1986 - Superfund Amendments and Reauthorization Act PL 99-499
- 2002 - Small Business Liability Relief and Brownfields Revitalization Act ("Brownfields Law") PL 107-118

Programs

Energy Star

In ,the EPA launched the Energy Star program, a voluntary program that fosters energy efficiency.

Pesticide Registration

EPA administers the Federal Insecticide, Fungicide, and Rodenticide Act (FIFRA) (which is much older than the Agency) and registers all pesticides legally sold in the United States.

Environmental Impact Statement Review

EPA is responsible for reviewing projects of other federal agencies' Environmental Impact Statements under the National Environmental Policy Act (NEPA).

Safer Detergents Stewardship Initiative

Through the Safer Detergents Stewardship Initiative (SDSI), EPA's Design for the Environment (DfE) Program recognizes environmental leaders who voluntarily commit to the use of safer surfactants. Safer surfactants are surfactants that break down quickly to non-polluting compounds and help protect aquatic life in both fresh and salt water. Nonylphenol ethoxylates, commonly referred to as NPEs, are an example of a surfactant class that does not meet the definition of a safer surfactant.

The Design for the Environment Program has identified safer alternative surfactants through partnerships with industry and environmental advocates. These safer alternatives are comparable in cost and are readily available. Clean Gredients is a source of safer surfactants.

Air Quality and Air Pollution

The Air Quality Modeling Group (AQMG) is in the EPA's Office of Air and Radiation (OAR) and provides leadership and direction on the full range of air quality models, air pollution dispersion models and other mathematical simulation techniques used in assessing pollution control strategies and the impacts of air pollution sources.

The AQMG serves as the focal point on air pollution modeling techniques for other EPA headquarters staff, EPA regional Offices, and State and local environmental agencies. It co-ordinates with the EPA's Office of Research and Development (ORD) on the development of new models and techniques, as well as wider issues of atmospheric research. Finally, the AQMG conducts modeling analyses to support the policy and regulatory decisions of the EPA's Office of Air Quality Planning and Standards (OAQPS).

Oil Pollution Prevention

SPCC - Spill Prevention Containment and Counter Measures. Secondary Containment mandated at oil storage facilities. Oil release containment at oil development sites.

Water Sense

The **Water Sense** program is designed to encourage water efficiency through the use of a special label on consumer products. Products include high-efficiency toilets

(HETs), bathroom sink faucets (and accessories), and irrigation equipment. Water Sense is a voluntary program, with EPA developing specifications for water-efficient products through a public process and product testing by independent laboratories. The program was launched in 2006.

Safe Drinking Water

EPA ensures safe drinking water for the public, by setting standards for more than 160,000 public water systems throughout the United States. EPA oversees states, local governments and water suppliers to enforce the standards under the Safe Drinking Water Act. The program includes regulation of injection wells in order to protect underground sources of drinking water.

Advance Identification

Advance Identification, or ADID, is a planning process used by the EPA to identify wetlands and other bodies of water and their respective suitability for the discharge of dredged and fill material. The EPA conducts the process in co-operation with the U.S. Army Corps of Engineers and local states or Native American Tribes. As of February 1993, 38 ADID projects had been completed and 33 were ongoing.

United States Government Agencies Involved in Environmental Science

United States Environmental Protection Agency	
Department of the Interior	National Park Service • Fish and Wildlife Service • Bureau of Indian Affairs • Bureau of Land Management • Minerals Management Service • Office of Surface Mining, Reclamation, and Enforcement • Geological Survey • Bureau of Reclamation • Office of Insular Affairs
National Oceanic and Atmospheric Administration	National Weather Service • National Ocean Service • National Geodetic Survey • National Oceanic and Atmospheric Administration • National Marine Fisheries Service • Office of Oceanic and Atmospheric Research
Department of Energy	Energy Efficiency and Renewable Energy • Energy Information Administration • Federal Energy Regulatory Commission • Biological and Environmental Research • Office of Environmental Management
Department of Agriculture	Farm Service Agency • Foreign Agricultural Service • Risk Management Agency • Food Safety and Inspection Service • Forest Service • Natural Resources Conservation Service • Rural Business-Cooperative Service • Office of Community Development • Rural Housing Service • Rural Utilities Service • Food and Nutrition Service • Center for Nutrition Policy and Promotion • Agricultural Marketing Service • Animal and Plant Health Inspection Service • Grain Inspection, Packers and Stockyards Administration • Agricultural Research Service • Co-operative State Research, Education, and Extension Service • Economic Research Service • National Agricultural Statistics Service • Agricultural Stabilization and Conservation Service

Chapter **17**

Metal-Ligand Interaction in Biological Fluids

Hydrolysis

Hydrolysis is a chemical reaction during which one or more water molecules are split into hydrogen and hydroxide ions which may go on to participate in further reactions. It is the type of reaction that is used to break down certain polymers, especially those made by step-growth polymerization. Such polymer degradation is usually catalysed by either acid, e.g., concentrated sulphuric acid [H_2SO_4] or alkali, e.g., sodium hydroxide [NaOH] attack, often increasing with their strength or pH.

Hydrolysis is distinct from hydration, where hydrated molecule does not "lyse" (break into two new compounds). It should not be confused with hydrogenolysis, a reaction of hydrogen.

Types

Hydrolysis is a chemical process in which a molecule is cleaved into two parts by the addition of a molecule of water. One fragment of the parent molecule gains a hydrogen ion (H+) from the additional water molecule. The other group collects the remaining hydroxyl group (OH-).

The most common hydrolysis occurs when a salt of a weak acid or weak base (or both) is dissolved in water. Water auto-ionizes into negative hydroxyl ions and hydrogen ions. The salt breaks down into positive and negative ions. For example, sodium acetate dissociates in water into sodium and acetate ions. Sodium ions react very little with hydroxyl ions whereas acetate ions combine with hydrogen ions to produce neutral acetic acid, and the net result is a relative excess of hydroxyl ions, causing a basic solution.

However, under normal conditions, only a few reactions between water and organic compounds occur. Generally, strong acids or bases must be added in order to achieve hydrolysis where water has no effect. The acid or base is considered a catalyst. They are meant to speed up the reaction, but are recovered at the end of it.

Acid–base catalyzed hydrolyses are very common; one example is the hydrolysis of amides or esters. Their hydrolysis occurs when the nucleophile (a nucleus-seeking agent, e.g., water or hydroxyl ion) attacks the carbon of the carbonyl group of the ester or amide. In an aqueous base, hydroxyl ions are better nucleophiles than dipoles such as water. In acid, the carbonyl group becomes protonated, and this leads to a much easier nucleophilic attack. The products for both hydrolyses are compounds with carboxylic acid groups.

Perhaps, the oldest example of ester hydrolysis is the process called saponification. It is the hydrolysis of a triglyceride (fat) with an aqueous base such as sodium hydroxide (NaOH). During the process, glycerol, also commercially named glycerin, is formed, and the fatty acids react with the base, converting them to salts. These salts are called soaps, commonly used in households.

Moreover, hydrolysis is an important process in plants and animals, the most significant example being energy metabolism and storage. All living cells require a continual supply of energy for two main purposes: for the biosynthesis of small macromolecules, and for the active transport of ions and molecules across cell membranes. The energy derived from the oxidation of nutrients is not used directly, but by means of a complex and long sequence of reactions, it is channeled into a special energy-storage molecule, adenosine triphosphate (ATP).

The ATP molecule contains pyrophosphate linkages (bonds formed when two phosphate units are combined together) that release energy when needed. ATP can be hydrolyzed in two ways: the removal of terminal phosphate to form adenosine diphosphate (ADP) and inorganic phosphate, or the removal of a terminal diphosphate to yield adenosine monophosphate (AMP) and pyrophosphate. The latter is usually cleaved further to yield two phosphates. This results in biosynthesis reactions, which do not occur alone, that can be driven in the direction of synthesis when the phosphate bonds are hydrolyzed.

In addition, in living-systems, most biochemical reactions, including ATP hydrolysis, take place during the catalysis of enzymes. The catalytic action of enzymes allows the hydrolysis of proteins, fats, oils, and carbohydrates. As an example, one may consider proteases, enzymes that aid digestion by hydrolyzing peptide bonds in proteins. They catalyze the hydrolysis of interior peptide bonds in peptide chains, as opposed to exopeptidases, another class of enzymes, that catalyze the hydrolysis of terminal peptide bonds, liberating one free amino acid at a time.

However, proteases do not catalyze the hydrolysis of all kinds of proteins. Their action is stereo-selective. Only proteins with a certain tertiary structure will be targeted. The reason is that some kind of orienting force is needed to place the amide group in the proper position for catalysis. The necessary contacts between an enzyme and its substrates (proteins) are created because the enzyme folds in such a way as to form a crevice into which the substrate fits; the crevice also contains the catalytic groups. Therefore, proteins that do not fit into the crevice will not be hydrolyzed. This specificity preserves the integrity of other proteins such as hormones, and therefore the biological system continues to function normally.

Hydrolysis of Amide Links

In the hydrolysis of an amide link into a carboxylic acid and an amine or ammonia, the carboxylic acid has a hydroxyl group derived from a water molecule and the amine (or ammonia) gains the hydrogen ion.

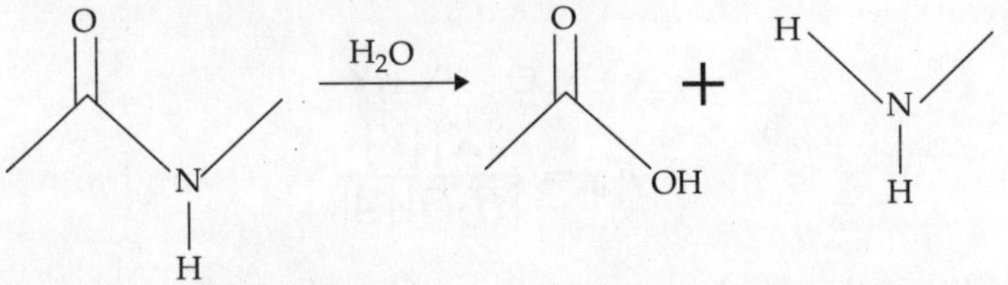

O, N, H — H_2O → O, OH + H, N, H

A specific case of the hydrolysis of an amide link is the hydrolysis of peptides to smaller fragments or amino acids.

Many polyamide polymers such as nylon 6,6 are attacked and hydrolysed in the presence of strong acids. Such attack leads to depolymerization and nylon products fail by fracturing when exposed to even small amounts of acid. Other polymers made by step-growth polymerization are susceptible to similar polymer degradation reactions. The problem is known as stress corrosion cracking.

Hydrolysis of Polysaccharides

CH_2OH, O, H, H, OH, H, OH, H, OH, O, CH_2OH, O, H, H, H_2O, CH_2OH, OH, H

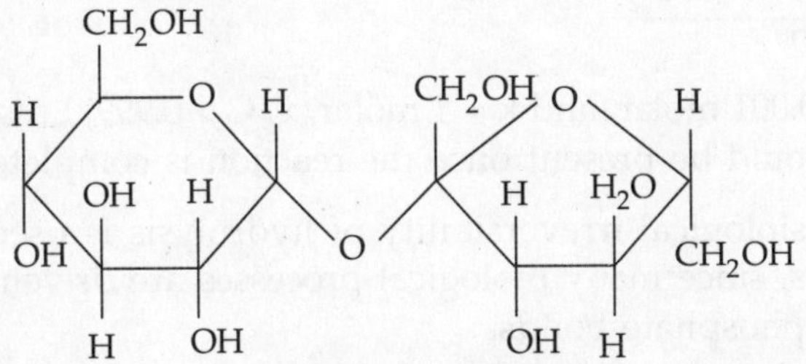

Sucrose. The glycoside bond is represented by the central oxygen atom which holds the two monosaccharide units together.

In polysaccharides, monosaccharide molecules are linked together by a glycosidic bond. This bond can be cleaved by hydrolysis to yield monosaccharides. The best known disaccharide is sucrose. Hydrolysis of sucrose yields glucose and fructose. There are many enzymes which speed up the hydrolysis of polysaccharides. Invertase is used industrially to hydrolyze sucrose to so-called invert sugar. Invertase is an example of a glycoside hydrolase (glucosidase). Lactase is essential for digestive hydrolysis of lactose in milk. Deficiency of the enzyme in humans causes lactose intolerance. β-amylase catalyzes the conversion of starch to maltose. Malt made from barley is used as a source of β-amylase to break down starch into a form that can be used by yeast to produce beer. The hydrolysis of cellulose into glucose, known as *saccharification*, is catalyzed by cellulase. Animals such as cows (ruminants) are able to digest cellulose because of the presence of symbiotic bacteria which produce cellulases.

Irreversibility of Hydrolysis under Physiological Conditions

Under physiological conditions (i.e., in dilute aqueous solution), a hydrolytic cleavage reaction, where the concentration of a metabolic precursor is low (of the order of 10^{-3} to 10^{-6} molar), is essentially thermodynamically irreversible. To give an example:

$$A + H_2O \rightarrow X + Y$$

$$K_d = \frac{[X]\,[Y]}{[H_2O]\,[A]}$$

Assuming that x is the final concentration of products, and that C is the initial concentration of A, and W = $[H_2O]$ = 55.5 molar, then x can be calculated with the equation:

$$\frac{x \times x}{W\,(C - x)} = K_d$$

let $K_d \times W = k$:

then $x = \dfrac{-k + \sqrt{k^2 + 4kC}}{2}$.

For a value of C = 0.001 molar, and k = 1 molar, $x/C > 0.999$. Less than 0.1% of the original reactant would be present once the reaction is complete.

This theme of physiological irreversibility of hydrolysis is used consistently in metabolic pathways, since many biological processes are driven by the cleavage of anhydrous pyrophosphate bonds.

Hydrolysis of Metal Aqua Ions

Metal ions are Lewis acids, and in aqueous solution they form aqua ions, of the general formula $M(H_2O)_n^{m+}$. The aqua ions are hydrolyzed, to a greater or lesser extent. The first hydrolysis step is given generically as:

$$M(H_2O)_n^{m+} + H_2O \rightleftharpoons M(H_2O)_{n-1}(OH)^{(m-1)+} + H_3O^+$$

Thus, the aqua ion is behaving as an acid in terms of Brønsted-Lowry acid-base theory. This is easily explained by considering the inductive effect of the positively charged metal ion, which weakens the O-H bond of an attached water molecule, making the liberation of a proton relatively easy.

The dissociation constant, pK_a, for this reaction is more or less linearly related to the charge-to-size ratio of the metal ion. Ions with low charges, such as Na^+ are very weak acids with almost imperceptible hydrolysis. Large divalent ions such as Ca^{2+}, Zn^{2+}, Sn^{2+} and Pb^{2+} have a pK_a of 6 or more and would not normally be classed as acids, but small divalent ions such as Be^{2+} are extensively hydrolyzed. Trivalent ions like Al^{3+} and Fe^{3+} are weak acids whose pK_a is comparable to that of acetic acid. Solutions of salts such as $BeCl_2$ or $Al(NO_3)_3$ in water are noticeably acidic; the hydrolysis can be suppressed by adding an acid such as nitric acid, making the solution more acidic.

Hydrolysis may proceed beyond the first step, often with the formation of polynuclear species. Some "exotic" species such as $Sn_3(OH)_4^{2+}$ are well characterized. Hydrolysis tends to increase as pH rises leading, in many cases, to the precipitation of an hydroxide such as $Al(OH)_3$ or $AlO(OH)$. These substances, the major constituents of bauxite, are known as laterites and are formed by leaching from rocks of most of the ions other than aluminium and iron and subsequent hydrolysis of the remaining aluminium and iron.

Ions with a formal charge of four are extensively hydrolyzed and salts of Zr^{4+}, for example, can only be obtained from strongly acidic solutions. With oxidation states five and higher the concentration of the aqua ion in solution is negligible. In effect the aqua ion is a strong acid. For example, aqueous solutions of Cr(VI) contain CrO_4^{2-}.

$$Cr(H_2O)_6^{6+} \rightarrow CrO_4^{2-} + 2\,H_2O + 8\,H^+$$

Note that reactions such as

$$Cr_2O_7^{2-} + H_2O \rightleftharpoons 2\,CrO_4^{2-} + 2\,H^+$$

are formally hydrolysis reactions as water molecules are split up yielding hydrogen ions. Such reactions are common among polyoxometalates.

METAL ION INTERACTION WITH MACROMOLECULES

Compounds of Zinc

Compounds of zinc are chemical compounds containing the element zinc which is a member of the group 12 of the periodic table. The oxidation state of most compounds is the group oxidation state of +2. Zinc may be classified as a post-transition main group element with zinc(II) having much chemical behaviour in common with copper(II). Many salts of zinc(II) are isomorphous with salts of magnesium(II) due to the ionic radii of the cations being almost the same. Zinc forms many complexes; metallo-proteins containing zinc are widespread in biological systems.

General Characteristics

Zinc atoms have an electronic configuration of $[Ar]3d^{10}4s^2$. When compounds in the +2 oxidation state are formed the *s* electrons are lost, so the bare zinc ion has the electronic configuration $[Ar]3d^{10}$. This allows for the formation of four covalent bonds by accepting four electron pairs and thus obeying the octet rule. The stereochemistry is tetrahedral and the bonds may be described as being formed from sp^3 hybrid orbitals on the zinc ion. Examples occur in the oxide, ZnO, (calamine) and sulphide, ZnS, (zinc blende) in which the oxide and sulphide ions are also tetrahedrally bound to four zinc ions. Many complexes, such as $ZnCl_4^{2-}$, are tetrahedral. Tetrahedrally co-ordinated zinc is found in metallo-enzymes such as carbonic anhydrase. However 6-coordinate complexes can also be formed by using empty 4d orbitals to form sp^3d^2 hybrid orbitals. The ion $[Zn(H_2O)_6]^{2+}$, which is present when a zinc salt is dissolved in water, has an octahedral structure.

Many zinc(II) salts are isomorphous (have the same type of crystal structure) with the corresponding salts of magnesium(II) which results from the fact that Zn^{2+} and Mg^{2+} have almost identical ionic radii. This comes about because of the d-block contraction. Whilst calcium is somewhat larger than magnesium, there is a steady decrease in size as atomic number increases from calcium to zinc. By chance it is the ionic radius of zinc that is almost equal to that of magnesium. In most other respects the chemistry of zinc(II) most closely resembles the chemistry of copper(II), its neighbour in the periodic table, in which there is less electron. However, whereas Cu^{2+} is classed as a transition metal ion by virtue of its electronic configuration, $[Ar]3d^9$, in which there is an incomplete *d*-shell, Zn^{2+} is best considered to be an ion of a post-transition main group element. The IUPAC periodic table places zinc in the d-block.

Some compounds with zinc in the oxidation state +1 are known. The compounds have the formula RZn_2R and they contain a Zn-Zn bond analogous to the metal-metal bond in mercury(I) ion, Hg_2^{2+}. In this respect, zinc is similar to magnesium where low valent compounds containing a Mg- Mg bond have been characterised.

No compounds of zinc in oxidation states other than +1 or +2 are known. Calculations indicate that a zinc compound with the oxidation state of +4 is unlikely to exist. Although higher oxidation states are more stable with the heavier elements of a group, the compound HgF_4 was only characterized at 4 K in an neon/argon matrix.

Colour and Magnetism

Zinc compounds, like those of main group elements, are mostly colourless. Exceptions occur when the compound contains a coloured anion or ligand. Zinc selenide, ZnSe, however, is yellow, due to charge-transfer transitions and zinc telluride, ZnTe is brown for the same reason. Zinc oxide turns yellow when heated due to the loss of some oxygen atoms and formation of a defective structure.

Compounds containing zinc and no other metal are all diamagnetic.

Reactivity of the Metal

Zinc is a strong reducing agent with a standard redox potential of -0.76 V. Pure zinc tarnishes rapidly in air, eventually forming a passive layer of basic zinc carbonate, $Zn_5(OH)_6CO_3$. The reaction of zinc with water is prevented by the passive layer. When this layer is penetrated by acids such as hydrochloric acid and sulfuric acid the reaction proceeds with the evolution of hydrogen gas.

$$Zn(s) + 2H^+ (aq) \rightarrow Zn^{2+} (aq) + H_2 \uparrow$$

The hydrogen ion is reduced by accepting an electron from the reducing agent. The zinc metal is oxidised. Amalgamation with mercury, as in the Jones reductor also destroys the passive layer. Zinc reacts with alkalis as with acids. It reacts directly with oxidising non-metals such as chalcogens and halogens to form binary compounds.

Binary Compounds

Zinc oxide, ZnO, is the most important manufactured compound of zinc, with a wide variety of uses. It crystallizes with the Wurtzite structure. It is amphoteric, dissolving in acids to give the aqueous Zn^{2+} ion and in alkali to give the tetrahedral hydroxo complex, $[Zn(OH)_4]^{2-}$. Zinc hydroxide, $Zn(OH)_2$ is also amphoteric.

Zinc sulphide, ZnS, crystallizes in two closely-related structures, the Zinc blende structure and the Wurtzite structure which are common structures of compounds with the formula MA. Both Zn and S are tetrahedrally co-ordinated by the other ion. A useful property of ZnS is its phosphorescence.The other chalcogenides, ZnSe, and ZnTe, have applications in electronics and optics.

Of the four halides ZnF_2 has the most ionic character, whereas the others, $ZnCl_2$, $ZnBr_2$, and ZnI_2, have relatively low melting points and are considered to have more covalent character. The pnictogenides Zn_3N_2 (notable for its high melting point), Zn_3P_2, Zn_3As_2 and Zn_3Sb_2, have various applications. Other binary compounds of zinc include the peroxide ZnO_2, the hydride ZnH_2, and the carbide ZnC_2.

Salts

The nitrate $Zn(NO_3)_2$ (used as oxidizing agent, the chlorate $Zn(ClO_3)_2$, the sulfate $ZnSO_4$ (known as "white vitriol"), the phosphate $Zn_3(PO_4)_2$ (used as primer pigment), the molybdate $ZnMoO_4$ (used as white pigment), the chromate $ZnCrO_4$ (one of the few colored zinc compounds), the arsenite $Zn(AsO_2)_2$ (colorless powder) and the arsenate octahydrate $Zn(AsO_4)_2 \cdot 8H_2O$ (white powder, also referred to as koettigite] are a few examples of other common inorganic compounds of zinc. The latter two compounds are both used in insecticides and wood preservatives. One of the simplest examples of an organic compound of zinc is the acetate $Zn(O_2CCH_3)_2$, which has several medicinal applications. Zinc salts are usually fully dissociated in aqueous solution. Exceptions occur when the anion can form a complex, such as in the case of zinc sulfate, where the complex $[Zn(H_2O)_n(SO_4)]$ may be formed, (log K = ca. 2.5).

Complexes

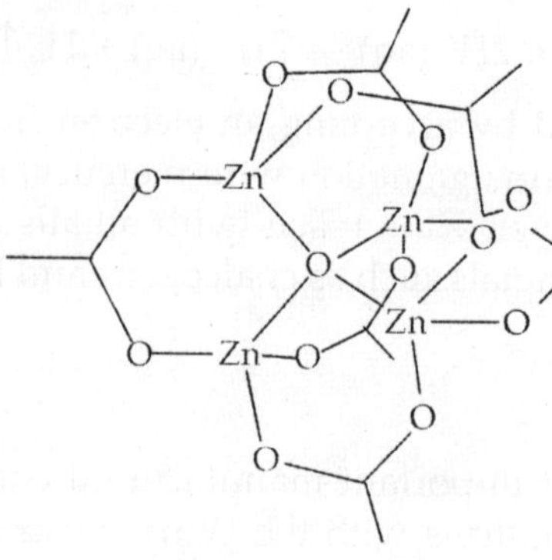

Structure of solid basic zinc acetate, $[Zn_4(m^4\text{-}O)(h^2\text{-}O_2CCH_3)_6]$

The most common structure of zinc complexes is tetrahedral which is clearly connected with the fact that the octet rule is obeyed in these cases. Nevertheless, octahedral complexes comparable to those of the transition elements are not rare. Zn^{2+} is a class. An acceptor in the classification of Ahrland, Chatt and Davies, and so forms stronger complexes with the first-row donor atoms oxygen or nitrogen than with second-row sulphur or phosphorus. In terms of HSAB theory Zn^{2+} is a hard acid.

In aqueous solution an octahedral complex, $[Zn(H_2O)_6]^{2+}$ is the predominant species. Aqueous solutions of zinc salts are mildly acidic because the aquaion is subject to hydrolysis with a pK_a of around 5, depending on conditions.

$$[Zn(H_2O)_6]^{2+} \rightleftharpoons [Zn(H_2O)_5(OH)]^{+} + H^{+}$$

Hydrolysis explains why basic salts such as basic zinc acetate and basic zinc carbonate, $Zn_3(OH)_4(CO_3).H_2O$ are easy to obtain. The reason for the hydrolysis is the high electrical charge density on the zinc ion, which pulls electrons away from an OH bond of a co-ordinated water molecule and releases a hydrogen ion. The polarizing effect of Zn^{2+} is part of the reason why zinc is found in enzymes such as carbonic anhydrase.

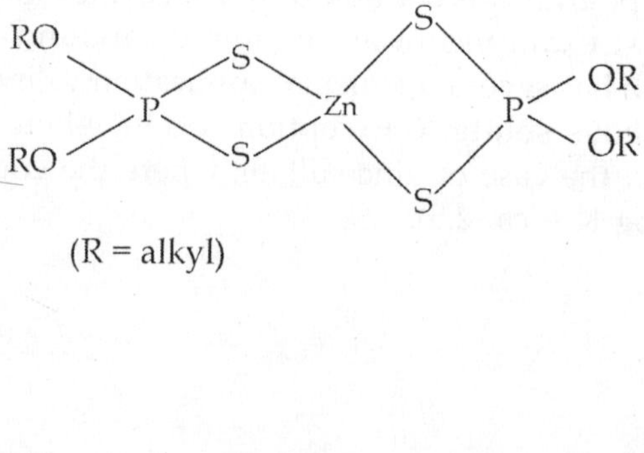

(R = alkyl)

Structure of a monomeric zinc dialkyldithiophosphate

No fluoro complexes are known, but complexes with the other halides and with pseodohalides, $[ZnX_3]^-$ and $[ZnX_4]$ can be prepared. The case of the thiocyanate complex illustrates the class A character of the zinc ion as it is the N-bonded isomer, $[Zn(NCS)_4]$ in contrast to $[Cd(SCN)_4]$ which is S-bonded. Being a class-A acceptor does not preclude the formation of complexes with sulphur donors, as is shown by Zinc dithiophosphate and the zinc finger complex (below).

The acetylacetonate complex, $Zn(acac)_2$ is interesting. As the ligand is bidentate a tetrahedral structure might be expected. However, the compound is in fact a trimer, $Zn_3(acac)_6$ in which each Zn ion is co-ordinated by five oxygen atoms in a distorted trigonal bipyramidal structure. Other 5-coordinate structures can be engineered by choosing ligands which have specific stereochemical requirements. For example, terpyridine, which is a tridentate ligand forms the complex $[Zn(terpy)Cl_2]$. Another example would involve a tripodal ligand such as Tris(2-aminoethyl)amine. The compound zinc cyanide, $Zn(CN)_2$, is not 2-coordinate. It adopts a polymeric structure consisting of tetrahedral zinc centres linked by bridging cyanide ligands. The cyanide group shows head to tail disorder with any zinc atom having between 1 and 4 carbon atom neighbours and the remaining being nitrogen atoms. These two examples illustrate the difficulty of sometimes relating structure to stoichiometry.

A co-ordination number of 2 occurs in the amide $Zn(NR^1R^2)_2$ (R^1=CMe_3, R^2=$SiMe_3$); the ligand is so bulky that there is not enough space for more than two of them.

Bio-complexes

A very large number of metallo-enzymes contain zinc(II). Also many proteins contain zinc for structural reasons. The zinc ion is invariably 4-coordinate with at least three ligands that are amino-acid side-chains. The imidazole nitrogen of a histidine side-chain is a common ligand. The following are typical examples of the two kinds of zinc-protein complexes.

In the active site of resting Carbonic anhydrase a zinc ion is co-ordinated by three histidine residues. The fourth position is occupied by a water molecule, which is strongly polarized as in hydrolysis (see above). When carbon dioxide enters the active site, it is subjected to nucleophilic attack by the oxygen atom which carries a partial negative charge, or indeed a full negative charge if the water molecule is dissociated. The CO_2 is rapidly converted into a bicarbonate ion.

$$[(\text{-hys})_3Zn(H_2O)]^{2+} + CO_2 \rightarrow [(\text{-hys})_3Zn]^{2+} + HCO_3^- + H^+$$

Some peptidases such as Glutamate carboxypeptidase II are thought to act in a similar way, with the zinc ion promoting the formation of a nucleophilic reagent.

The zinc finger motif is a rigid substructure in a protein which facilitates the binding of the protein to another molecule such as DNA. In this case, all four coordination positions are occupied by the hystidine and cysteine residues. The tetrahedral geometry around the zinc ion constrains an α helix fragment and an antiparallel β sheet fragment to a particular orientation with respect to each other.

The magnesium ion has a higher concentration in biological fluids, cannot perform these functions as its complexes are much weaker than those of zinc.

Organometallic Compounds

$$\begin{array}{l} H_3C \\ \quad \backslash \\ \quad H_2C — Zn — CH_2 \\ \qquad\qquad\qquad \backslash \\ \qquad\qquad\qquad CH_3 \end{array}$$

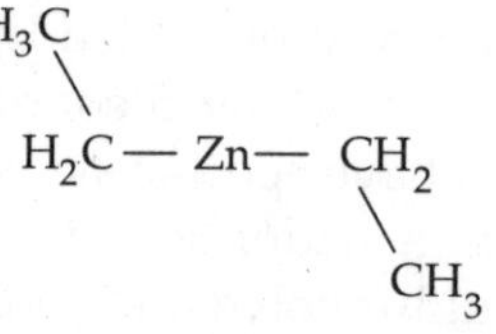

Diethylzinc

Organozinc compounds contain zinc-carbon covalent bonds. Diethylzinc ($(C_2H_5)_2Zn$) was first reported in 1848. It was made by reaction of zinc and ethyl iodide and is the first compound known to contain a metal-carbon sigma bond. For a long time it was a mystery why copper(II) did not form an analogous compound. It was not until the 1980s that the reason was found: the zinc compound does not undergo the Beta-hydride elimination reaction whereas the compound of the transition metal copper does so. Alkyl and aryl zinc compounds are contain the linear C—Zn—C motif as the zinc centre is co-ordinatively unsaturated and the compounds are powerful electrophiles. In fact the low-molecular weight compounds will ignite spontaneously on contact with air and are immediately destroyed by reaction with water molecules. The use of zinc alkyls has been largely superseded by the use of the more easily handled Grignard reagents. This demonstrates yet another connection between the chemistries of zinc and magnesium.

Zinc cyanide, $Zn(CN)_2$, is used as a catalyst in some organic reactions.

Organometallic compounds of zinc(I) contain M- M bonds. Decamethyldizincocene is now known.

METAL PROTEIN INTERACTION

Macrocycle

A **macrocycle** is, as defined by IUPAC, *"a cyclic macromolecule or a macromolecular cyclic portion of a molecule."* In the chemical literature, organic chemists may consider any molecule containing a ring of seven or more atoms to be a macrocylce. Co-ordination chemists generally define a macrocycle more narrowly as a cyclic molecule with three or more potential donor atoms that can co-ordinate to a metal center.

Macrocycle Effect

The macrocyclic effect was discovered in 1969. Co-ordination chemists study macrocycles with three or more potential donor atoms in rings of greater than nine atoms as these compounds often have strong and specific binding with metals. This property of co-ordinating macrocyclic molecules is the macrocycle effect. It is in essence a specific case of the chelation effect: complexes of bidentate and polydentate ligands are more stable than those with unidentate ligands of similar strength (or similar donor atoms). A macrocycle has donor atoms arranged in more fixed positions, and thus, there is less of an entropic effect in the binding energy of macrocycles than monodentate or bidentate ligands with an equal number of donor atoms. Thus the macrocycle effect states that complexes of macrocyclic ligands are more stable than those with linear polydentate ligands of similar strength (or similar donor atoms). The same can be said for multicyclic macrocycles, or cryptates, being stronger complexing agents (a cryptate effect).

Synthesis

Macrocycles are generally synthesized from smaller, usually linear, molecules. To create a ring, either an intermolecular reaction, where two or more molecules come together in a reaction to form a ring, or an intramolecular reaction, where one molecule reacts with itself to form a ring, must occur. Because the formation of macrocycles uses the same chemistry that polymerization does, steps need to be taken to prevent polymerization from occurring. Traditionally, this involved high dilution chemistry where large amounts of solvent and low concentrations were used to prevent molecules from reacting with other molecules. Also, the reagents frequently needed to be added slowly. At low concentration, the molecule is more likely to react with itself than with another molecule. This is generally inefficient, using large quantities of solvents and giving low yields.

To achieve high yields of macrocycles at high concentrations, a way to orient the reactive sites such that they readily undergo cyclization was needed. Transition metals, with their ability to gather and dispose off ligands in a given predictable geometry, can induce a "template effect." By binding to the linear molecule, to influence its geometry, a metal "template" can accelerate either the intramolecular or the intermolecular reaction. Thus the judicious choice of a metal ion and the relative locations of donor atoms would allow a metal to control the cyclization process.

The template effect can be divided into two slightly more specific effects: The kinetic template effect describes the directive influence of the metal ion controls the steric course of a sequence of step wise reactions. In cases where the thermodynamic template effect operates, the metal ion perturbs an existing equilibrium in an organic system and the required product is produced often in high yield as a metal complex. In most cases, the kinetic template effect is operative, however an assignment cannot be made in all cases.

Applications

- Removal of heavy metals from aqueous solution for water purification
  - Chelation therapy whereby the use of chelating agents such as EDTA to remove heavy metals from the body
- Molecular switches and linear motors for constructing artificial nanoscale machinery (rotaxanes)
- Chemical Sensors
- Mimicry of cellular receptors
- Molecular recognition
  - Recognition of peptides
  - Small molecules
- Organic light-emitting diodes (OLEDs)

Historical Uses

Macrocycles have been in use for several decades as synthetic dyes. Phthalocyanine is a porphyrin analogue, which is arguably the most useful, in uses as dyes and pigments since their discovery in 1928, due to their dark blue colour. There are however many other uses for them. Their name comes from their synthetic precursor, phthalodinitrile.

Biological Macrocycles

- Heme, the active site in the hemoglobin (the protein in blood that transports oxygen) is a porphyrin containing iron.
- Chlorophyll, the green photosynthetic pigment found in plants contains a chlorin ring.
- Vitamin B12, contains a corrin ring.

Related Molecular Categories

- Ligand: an atom, ion or functional group that is bonded to one or more central atoms or ions.
- Chelate: a multidentate ligand, containing more than one donor atom.
- Cryptand: a macrocycle with multiple loops (e.g. bicyclic).
- Rotaxane: macrocycle(s) stuck on a stick, generally freely.
- Catenane: interlocked molecular rings (like a chain).
- Molecular knot: a molecule in the shape of a knot such as a trefoil knot.

METAL NUCLEIC ACID INTERACTION

Nucleic Acid Analogues

Nucleic acid analogues are compounds structurally similar (analog) to naturally occurring RNA and DNA, used in medicine and in molecular biology research.

Nucleic acids are chains of nucleotides, which are composed of three parts: a phosphate backbone, a pucker-shaped pentose sugar, either ribose or deoxyribose, and one of four nucleobases. An analogue may have any of these altered. Typically the analogue nucleobases confer, among other things, different base pairing and base stacking properties. Examples include universal bases, which can pair with all four canon bases, and phosphate-sugar backbone analogues such as PNA, which affect the properties of the chain (PNA can even form a triple helix).

Medicine

Several nucleoside analogues are used as antiviral or anticancer agents. The viral polymerase incorporates these compounds with non-canon bases. These compounds are activated in the cells by being converted into nucleotides, they are administered as nucleosides since charged nucleotides cannot easily cross cell membranes.

Common Changes in Nucleotide Analogues

Nucleic acid analogues are used in molecular biology for several purposes:

- As a tool to detect particular sequences
- As a tool with resistance to RNA hydrolysis
- As a tool for another purpose, such as sequencing
- Naturally occurring, such as in tRNA
- Investigation of the mechanisms used by enzyme, such as an Enzyme Inhibitor
- Investigation of possible scenarios of the origin of life
- Investigation of the structural features of nucleic acids
- Investigation of the possible alternatives to the natural system in Synthetic Biology

Backbone Analogues

Hydrolysis resistant RNA-analogues

OH
O Base
N
$H_2N-P=O$
O

Chemical Structure of Morpholino

To overcome the fact that ribose's 2' hydroxy group that reacts with the phosphate linked 3' hydroxy group (RNA is too unstable to be used or synthesised reliably), a ribose anologue is used. The most common RNA analogues are locked nucleic acid (LNA), morpholino, and peptide nucleic acid (PNA). These oligonucleotides differ as they have a different backbone sugar but still bind according to Watson and Crick pairing with RNA or DNA, but are immune to nuclease activity. (They generally cannot be enzymatically synthesized and can only be produced synthetically.)

Other notable analogues used as tools:

Dideoxynucleotides are used in sequencing. These nucleotide triphosphates possess a non-canonical sugar, dideoxyribose, which lacks the 3' hydroxyl group normally present in DNA and therefore cannot bond with the next base. The lack of the 3' hydroxyl group terminates the chain reaction as the DNA polymerases mistake it for a regular deoxyribonucleotide. The nucleoside analogue with a ribose lacking both 2' and 3' is called cordycepin. Cordycepin is an anticancer drug. Another analogue in sequencing is a nucleobase analogue, 7-deaza-GTP and is used to sequence CG rich regions, instead 7-deaza-ATP is called tubercidin, an antibiotic.

Precursors to the RNA World:

RNA may be too complex to be the first nucleic acid, so before the RNA world several simpler nucleic acids which differ in the backbone, such as TNA and GNA and PNA, have been offered as candidates for the first nucleic acids.

Base Analogues

Nucleobase Structure and Nomenclature

Natural bases are divided into two classes depending on their structure: pyrimidine (an heterocyclic aromatic six-membered ring with nitrogen atoms in position 1 and 3) and purine (a pyrimidine (numeration inverted) fused with an imidazole ring), a five-membered ring with 2 nitrogen atoms separated by one carbon (meta), 7,9. Their main proprieties are base pairing, resulting form 2 or 3 hydrogen bonds between keto and amino functional groups, and base stacking, caused by the attraction of the delocalized pi electron clouds of the aromatic ring structure.

Purine Pyrimidine

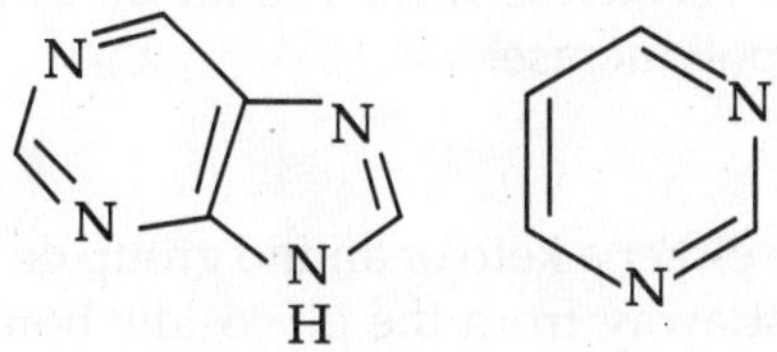

Fluorophores

Fluorophore

Structure of aminoallyl-uridine

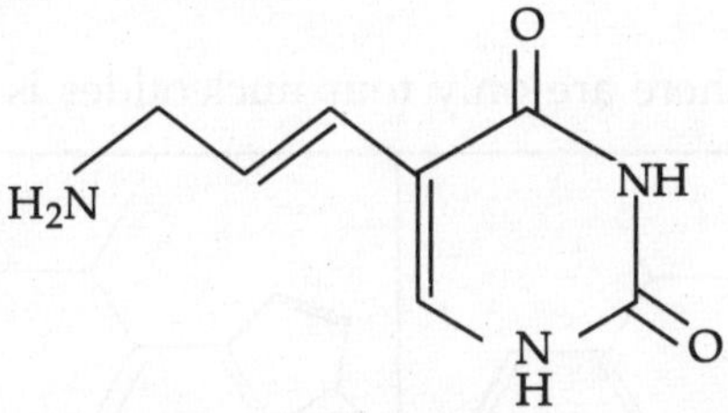

Commonly fluorophores (such as rhodamine or fluorescein) are linked to the ring which is linked to the sugar (in para) via a flexible arm, presumably extruding from the major groove of the helix. Due to low processivity of the nucleotides linked to bulky adducts such as florophores by taq polymerases, the sequence is typically copied using a nucleotide with an arm and later coupled with a reactive fluorophore (indirect labeling):

- Amine reactive: Aminoallyl nucleotide contain a primary amine group on a linker that reacts with the amino-reactive dye such as a cyanine or Alexa Fluor dyes, which contain a reactive leaving group, such as a succinimidyl ester (NHS). (Base pairing amino groups are not affected.)
- Thiol reactive: Thiol containing nucleotides reacts with the fluorophore linked to a reactive leaving group, such as a maleimide.
- Biotin-linked nucleotides rely on the same indirect labelling principle (+ fluorescent streptavidin) and are used in Affymetrix DNAchips.

Fluorophores find a variety of uses in medicine and biochemistry.

Natural Non-canon Bases

In a cell, there are several non-canon bases present: CpG islands in DNA (are often methylated), all eukaryotic mRNA (capped with a methyl-7-guanosine), and several bases of rRNAs (are methylated). Often, tRNAs are heavily modified postranscriptionally in order to improve their conformation or base pairing in particular in/near the anticodon: inosine can base pair with C, U, and even with A, whereas thiouridine (with A) is more specific than uracil (with a purine). Other common tRNA base modifications are pseudorindine (which gives its name to the TψC loop), dihydrouridine (which does not stack as it is not aromatic), queosine, wyosine and so forth. Nevertheless these are all modifications to normal bases and are not placed by a polymerase.

Base-pairing

Canonical bases may have either a keto or amino group on the carbons surrounding the nitrogen atom furthest away from the glycosidic bond, which allows them to base pair (Watson-Crick base pairing) via hydrogen bonds (amine with keto, purine with pyrimidine). A and T are amine only and keto only, while C and G are mixed (inverted in respect to each other).

Natural Base Pairs

The precise reason why there are only four nucleotides is debated, but there are

Guanine Cytosine	Adenine Thymine
A GC ba sepair: purine keto/amine forms three intermolecular hydrogen bonds with pyrimidine amine/keto	An AT basepair: purine amine/- forms two intermolecular hydrogen bonds with pyrimidine keto/keto

several unused possibilities. Furthermore adenine is not the most stable choice for base-pairing: in Cyanophage S-2L diaminopurine (DAP) is used instead of adenine (host evasion). Diaminopurine basepairs perfectly with thymine as it is identical to adenine but has an amine group at position 2 forming 3 intramolecular hydrogen bonds, eliminating the major difference between the two types of basepairs (Weak: A-T and Strong: C-G). This improved stability affects protein binding ineractions which rely on those differences. Other combinations include:

- isoguanosine and isocytosine, which are have their amino and keto inverted, (not used probably as tautomers are problematic for base-pairing, but isoG and isoG can be amplified correctly with PCR even in the presence of the 4 canon bases).
- diaminopyrimidine and xanthine (not used as xanthine is a deamination product).

Unused Base Pair Arrangements

Diaminopurine / Thymine	Xanthine / Diaminopyrimidine	Isoguanine / Isocytosine
A DAP -T base: purine amine/amine forms three intermolecular hydrogen bonds with pyrimidine keto/keto	An X -DAY base: purine keto/keto forms three intermolecular hydrogen bonds with pyrimidine amine/amine	A iG -iC base: purine amine/keto forms three intermolecular hydrogen bonds with pyrimidine keto/amine

However correct DNA structure can form even when the bases are not paired via hydrogen bonding, as studies have shown using DNA isosteres (analogues with same number of atoms), such as the thymine analogue 2,4-difluorotoluene (F) or the adenine analogue 4-methylbenzimidazole (Z).

Other noteworthy base pairs:

- Several fluorescent bases have also been made, such as the 2-amino-6-(2-thienyl)purine and pyrrole-2-carbaldehyde base pair.
- Metal co-ordinated bases, such as two 2,6-bis(ethylthiomethyl)pyridine (SPy) with a silver ion or pyridine-2,6-dicarboxamide (Dipam) and a mondentate pyridine (Py) with a copper ion.
- Universal bases may pair indiscriminately with any other base, but generally lower the melting temperature of the sequence considerably, examples include 2'-deoxyinosine derivatives, nitroazole analogues and hydrophobic aromatic non-hydrogen bonding bases (strong stacking effects). These are used as proof of concept and are not generally utilised in degenerate primers (which are a mixture of primers).
- The numbers of possible base pairs is doubled when xDNA is considered. xDNA contains expanded bases, in which a benzene ring has been added, which may pair with canon bases, resulting in four possible base-pairs (8 bases: xA-T,xT-A,xC-G,xG-C, 16 bases if the unused arrangements are used). Another form of benzene added bases is yDNA, in which the base is widened by the benzene.

Metal Base Pairs

In metal base-pairing, the Watson-Crick hydrogen bonds are replaced by the interaction between a metal ion with nucleosides acting as ligands. The possible geometries of the metal that would allow for duplex formation with two bidentate nucleosides around a central metal atom are: tetrahedral, dodecahedral, and square planar. Metal-complexing with DNA can occur by the formation of non-canonical base pairs from natural nucleobases with participation by metal ions and also by the exchanging the hydrogen atoms that are part of the Watson-Crick base pairing by metal ions. Introduction of metal ions into a DNA duplex has shown to have potential magnetic, conducting properties, as well as increased stability.

Metal complexing has been shown to occur between natural nucleobases. A well documented example is the formation of T-Hg-T, which involves two deprotonated thymine nucleobases that are brought together by Hg^{2+} and forms a connected metal-base pair. This motif does not accommodate stacked Hg^{2+} in a duplex, due to an intrastrand hairpin formation process that is favored over duplex formation. Two thymines across from each other in a duplex do not form a Watson-Crick base pair in a duplex; this is an example where a Watson-Crick base pair mismatch is stabilized by the formation of the metal-base pair. Another example of a metal complexing to natural nucleobases is the formation of A-Zn-T and G-Zn-C at high pH; Co^{+2} and Ni^{+2} also form these complexes. These are Watson-Crick base pairs where the divalent cation in co-ordinated to the nucleobases. The exact binding is debated.

A large variety of artificial nucleobases have been developed for use as metal base pairs. These modified nucleobases exhibit tunable electronic properties, sizes, and binding affinities that can be optimized for a specific metal. For example, a nucleoside modified with a pyridine-2,6-dicarboxylate has shown to bind tightly to Cu^{2+}, whereas other divalent ions are only loosely bound. The tridentate character contributes to this selectivity. The fourth co-ordination site on the copper is saturated by an oppositely arranged pyridine nucleobase. The asymmetric metal base pairing system is orthogonal to the Watson-Crick base pairs. Another example of an artificial nucleobase is that with hydroxypyridone nucleobases, which are able to bind Cu^{2+} inside the DNA duplex. Five consecutive copper-hydroxypyridone base pairs were incorporated into a double strand, which were flanked by only one natural nucleobase on both ends. EPR data showed that the distance between copper centers was estimated to be 3.7 ± 0.1 Å, while a natural B-type DNA duplex is only slighly larger (3.4 Å). The appeal for stacking metal ions inside a DNA duplex is the hope to obtain nanoscopic self-assembling metal wires, though this has not been realized yet.

RNA

Ribonucleic acid (RNA) is a type of molecule that consists of a long chain of nucleotide units. Each nucleotide consists of a nitrogenous base, a ribose sugar, and a phosphate. RNA is very similar to DNA, but differs in a few important structural details: in the cell, RNA is usually single-stranded, while DNA is usually double-stranded; RNA nucleotides contain ribose while DNA contains deoxyribose (a type of ribose that lacks one oxygen atom); and RNA has the base uracil rather than thymine that is present in DNA.

RNA is transcribed from DNA by enzymes called RNA polymerases and is generally further processed by other enzymes. RNA is central to the synthesis of proteins. Here, a type of RNA called messenger RNA carries information from DNA to structures called ribosomes. These ribosomes are made from proteins and ribosomal RNAs, which come together to form a molecular machine that can read messenger RNAs and translate the information they carry into proteins. There are many RNAs with other roles – in particular regulating which genes are expressed, but also as the genomes of most viruses.

Each nucleotide in RNA contains a ribose sugar, with carbons numbered 1' through 5'. A base is attached to the 1' position, generally adenine (A), cytosine (C), guanine (G) or uracil (U). Adenine and guanine are purines, cytosine and uracil are pyrimidines. A phosphate group is attached to the 3' position of one ribose and the 5' position of the next. The phosphate groups have a negative charge each at physiological pH, making RNA a charged molecule (polyanion). The bases may form hydrogen bonds between cytosine and guanine, between adenine and uracil and between guanine and uracil. However, other interactions are possible, such as a group of adenine bases binding to each other in a bulge, or the GNRA tetraloop that has a guanine–adenine base pair.

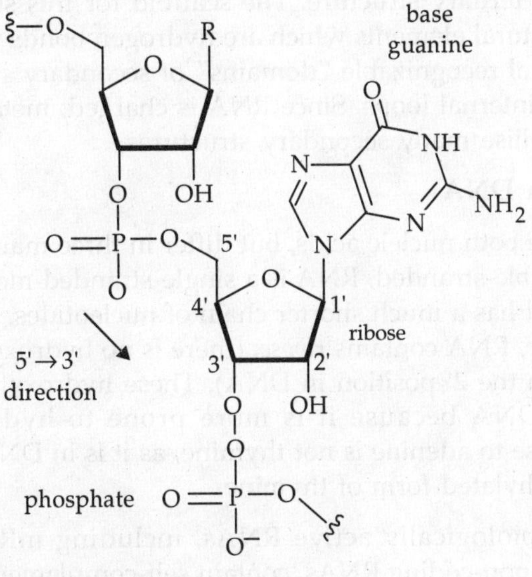

Chemical Structure of RNA

An important structural feature of RNA that distinguishes it from DNA is the presence of a hydroxyl group at the 2' position of the ribose sugar. The presence of this functional group causes the helix to adopt the A-form geometry rather than the B-form most commonly observed in DNA. This results in a very deep and narrow major groove and a shallow and wide minor groove. A second consequence of the presence of the 2'-hydroxyl group is that in conformationally flexible regions of an RNA molecule (that is, not involved in formation of a double helix), it can chemically attack the adjacent phosphodiester bond to cleave the backbone.

RNA is transcribed with only four bases (adenine, cytosine, guanine and uracil), but there are numerous modified bases and sugars in mature RNAs. Pseudouridine (ψ), in which the linkage between uracil and ribose is changed from a C–N bond to a C–C bond, and ribothymidine (T), are found in various places (most notably in the TψC loop of tRNA). Another notable modified base is hypoxanthine, a deaminated adenine base whose nucleoside is called inosine (I). Inosine plays a key role in the wobble hypothesis of the genetic code. There are nearly 100 other naturally occurring modified nucleosides, of which pseudouridine and nucleosides with 2'-O-methylribose are the most common. The specific roles of many of these modifications in RNA are not fully understood. However, it is notable that in ribosomal RNA, many of the post-transcriptional modifications occur in highly functional regions, such as the peptidyl transferase center and the subunit interface, implying that they are important for normal function.

The functional form of single stranded RNA molecules, just like proteins, frequently requires a specific tertiary structure. The scaffold for this structure is provided by secondary structural elements which are hydrogen bonds within the molecule. This leads to several recognizable "domains" of secondary structure like hairpin loops, bulges and internal loops. Since RNA is charged, metal ions such as Mg^{2+} are needed to stabilise many secondary structures.

Comparison with DNA

RNA and DNA are both nucleic acids, but differ in three main ways. First, unlike DNA which is double-stranded, RNA is a single-stranded molecule in most of its biological roles and has a much shorter chain of nucleotides. Second, while DNA contains *deoxyribose*, RNA contains *ribose*, (there is no hydroxyl group attached to the pentose ring in the 2' position in DNA). These hydroxyl groups make RNA less stable than DNA because it is more prone to hydrolysis. Third, the complementary base to adenine is not thymine, as it is in DNA, but rather uracil, which is an unmethylated form of thymine.

Like DNA, most biologically active RNAs, including mRNA, tRNA, rRNA, snRNAs and other non-coding RNAs, contain self-complementary sequences that allow parts of the RNA to fold and pair with itself to form double helices. Structural analysis of these RNAs have revealed that they are highly structured. Unlike DNA, their structures do not consist of long double helices but rather collections of short helices packed together into structures akin to proteins. In this fashion, RNAs can achieve chemical catalysis, like enzymes. For instance, determination of the structure of the ribosome—an enzyme that catalyzes peptide bond formation—revealed that its active site is composed entirely of RNA.

Synthesis

Synthesis of RNA is usually catalyzed by an enzyme—RNA polymerase—using DNA as a template, a process known as transcription. Initiation of transcription begins with the binding of the enzyme to a promoter sequence in the DNA (usually found "upstream" of a gene). The DNA double helix is unwound by the helicase activity of the enzyme. The enzyme then progresses along the template strand in the 3′ to 5′ direction, synthesizing a complementary RNA molecule with elongation occurring in the 5′ to 3′ direction. The DNA sequence also dictates where termination of RNA synthesis will occur.

RNAs are often modified by enzymes after transcription. For example, a poly(A) tail and a 5' cap are added to eukaryotic pre-mRNA and introns are removed by the spliceosome.

There are also a number of RNA-dependent RNA polymerases that use RNA as their template for synthesis of a new strand of RNA. For instance, a number of RNA viruses (such as poliovirus) use this type of enzyme to replicate their genetic material. Also, RNA-dependent RNA polymerase is part of the RNA interference pathway in many organisms.

Types of RNA

Messenger RNA

Messenger RNA (mRNA) is the RNA that carries information from DNA to the ribosome, the sites of protein synthesis (translation) in the cell. The coding sequence of the mRNA determines the amino acid sequence in the protein that is produced. Many RNAs do not code for protein, however, (about 97% of the transcriptial output is non-protein-coding in eukaryotes).

These so-called non-coding RNAs ("ncRNA") can be encoded by their own genes (RNA genes), but can also derive from mRNA introns. The most prominent examples of non-coding RNAs are transfer RNA (tRNA) and ribosomal RNA (rRNA), both of which are involved in the process of translation. There are also non-coding RNAs involved in gene regulation, RNA processing and other roles, possibly including a postulated role as the physical embodiment of Piaget's abstract "schèmes" which he saw as underlying repeatable action patterns and advanced thought-and-memory. Certain RNAs are able to catalyse chemical reactions such as cutting and ligating other RNA molecules, and the catalysis of peptide bond formation in the ribosome; these are known as ribozymes.

In-translation

Messenger RNA (mRNA) carries information about a protein sequence to the ribosomes, the protein synthesis factories in the cell. It is coded so that every three nucleotides (a codon) correspond to one amino acid. In eukaryotic cells, once precursor mRNA (pre-mRNA) has been transcribed from DNA, it is processed to mature mRNA. This removes its introns—non-coding sections of the pre-mRNA. The mRNA is then exported from the nucleus to the cytoplasm, where it is bound to ribosomes and translated into its corresponding protein form with the help of tRNA. In prokaryotic cells, which do not have nucleus and cytoplasm compartments, mRNA can bind to ribosomes while it is being transcribed from DNA. After a certain amount of time the message degrades into its component nucleotides with the assistance of ribonucleases.

Transfer RNA (tRNA) is a small RNA chain of about 80 nucleotides that transfers a specific amino acid to a growing polypeptide chain at the ribosomal site of protein synthesis during translation. It has sites for amino acid attachment and an anticodon region for codon recognition that binds to a specific sequence on the messenger RNA chain through hydrogen bonding.

Ribosomal RNA (rRNA) is the catalytic component of the ribosomes. Eukaryotic ribosomes contain four different rRNA molecules: 18S, 5.8S, 28S and 5S rRNA. Three of the rRNA molecules are synthesized in the nucleolus, and one is synthesized elsewhere. In the cytoplasm, ribosomal RNA and protein combine to form a nucleoprotein called a ribosome. The ribosome binds mRNA and carries out protein synthesis. Several ribosomes may be attached to a single mRNA at any time. rRNA is extremely abundant and makes up 80% of the 10 mg/ml RNA found in a typical eukaryotic cytoplasm.

Transfer-messenger RNA (tmRNA) is found in many bacteria and plastids. It tags proteins encoded by mRNAs that lack stop codons for degradation and prevents the ribosome from stalling.

Regulatory RNAs

Several types of RNA can downregulate gene expression by being complementary to a part of an mRNA or a gene's DNA. MicroRNAs (miRNA; 21-22 nt) are found in eukaryotes and act through RNA interference (RNAi), where an effector complex of miRNA and enzymes can break down mRNA which the miRNA is complementary to block the mRNA from being translated, or accelerate its degradation. While small interfering RNAs (siRNA; 20-25 nt) are often produced by breakdown of viral RNA, there are also endogenous sources of siRNAs. siRNAs act through RNA interference in a fashion similar to miRNAs. Some miRNAs and siRNAs can cause genes they target to be methylated, thereby decreasing or increasing transcription of those genes. Animals have Piwi-interacting RNAs (piRNA; 29-30 nt) which are active in germline cells and are thought to be a defense against transposons and play a role in gametogenesis. All prokaryotes have CRISPR RNAs, a regulatory system analogous to RNA interference. Antisense RNAs are widespread; most downregulate a gene, but a few are activators of transcription. One way antisense RNA can act is by binding to an mRNA, forming double-stranded RNA that is enzymatically degraded. There are many long noncoding RNAs that regulate genes in eukaryotes, one such RNA is Xist which coats one X chromosome in female mammals and inactivates it.

An mRNA may contain regulatory elements itself, such as riboswitches, in the 5' untranslated region or 3' untranslated region; these cis-regulatory elements regulate the activity of that mRNA. The untranslated regions can also contain elements that regulate other genes.

In RNA Processing

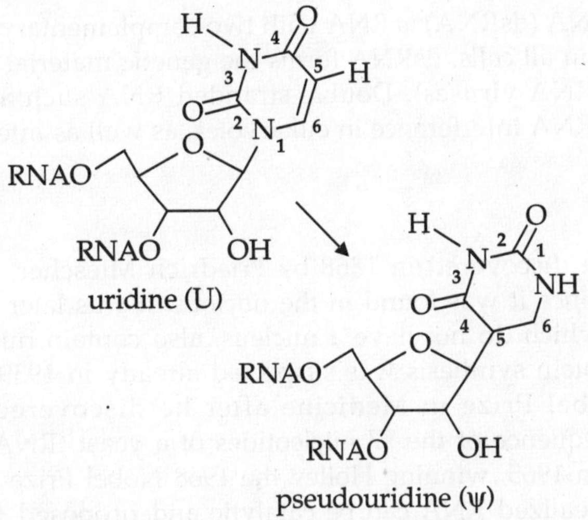

uridine (U)

pseudouridine (ψ)

Uridine to pseudouridine is a common RNA modification.

Many RNAs are involved in modifying other RNAs. Introns are spliced out of pre-mRNA by spliceosomes, which contain several small nuclear RNAs (snRNA), or the introns can be ribozymes that are spliced by themselves. RNA can also be altered by having its nucleotides modified to other nucleotides than A, C, G and U. In eukaryotes, modifications of RNA nucleotides are generally directed by small nucleolar RNAs (snoRNA; 60-300 nt), found in the nucleolus and cajal bodies. snoRNAs associate with enzymes and guide them to a spot on an RNA by basepairing to that RNA. These enzymes then perform the nucleotide modification. rRNAs and tRNAs are extensively modified, but snRNAs and mRNAs can also be the target of base modification.

RNA genomes

Like DNA, RNA can carry genetic information. RNA viruses have genomes composed of RNA, and a variety of proteins encoded by that genome. The viral genome is replicated by some of those proteins, while other proteins protect the genome as the virus particle moves to a new host cell. Viroids are another group of pathogens, but they consist only of RNA, do not encode any protein and are replicated by a host plant cell's polymerase.

In Reverse Transcription

Reverse transcribing viruses replicate their genomes by reverse transcribing DNA copies from their RNA; these DNA copies are then transcribed to new RNA. Retrotransposons also spread by copying DNA and RNA from one another, and telomerase contains an RNA that is used as template for building the ends of eukaryotic chromosomes.

Double-stranded RNA

Double-stranded RNA (dsRNA) is RNA with two complementary strands, similar to the DNA found in all cells. dsRNA forms the genetic material of some viruses (double-stranded RNA viruses). Double-stranded RNA such as viral RNA or siRNA can trigger RNA interference in eukaryotes, as well as interferon response in vertebrates.

Discovery

Nucleic acids were discovered in 1868 by Friedrich Miescher, who called the material 'nuclein' since it was found in the nucleus. It was later discovered that prokaryotic cells, which do not have a nucleus, also contain nucleic acids. The role of RNA in protein synthesis was suspected already in 1939. Severo Ochoa won the 1959 Nobel Prize in Medicine after he discovered how RNA is synthesized. The sequence of the 77 nucleotides of a yeast tRNA was found by Robert W. Holley in 1965, winning Holley the 1968 Nobel Prize in Medicine. In 1967, Carl Woese realized RNA can be catalytic and proposed that the earliest forms of life relied on RNA both to carry genetic information and to catalyze biochemical reactions—an RNA world. In 1976, Walter Fiers and his team determined the first complete nucleotide sequence of an RNA virus genome, that of bacteriophage MS2. In 1990 it was found in petunia that introduced genes can silence similar genes of the plant's own, now known to be a result of RNA interference. At about the same time, 22 nt long RNAs, now called microRNAs, were found to have a role in the development of *C. elegans*. The discovery of gene regulatory RNAs has led to attempts to develop drugs made of RNA, such as siRNA, to silence genes.

Chapter 18

Environmental Impact Assessment

An environmental impact assessment (EIA) is an assessment of the possible impact—positive or negative—that a proposed project may have on the environment; impact assessment considering natural, social and economic aspects. The purpose of the assessment is to ensure that decision makers consider the ensuing environmental impacts to decide whether to proceed with the project. The International Association for Impact Assessment (IAIA) defines an environmental impact assessment as "the process of identifying, predicting, evaluating and mitigating the biophysical, social, and other relevant effects of development proposals prior to major decisions being taken and commitments made." After an EIA, the precautionary and polluter pays principles may be applied to prevent, limit, or require strict liability or insurance coverage to a project, based on its likely harms. Environmental impact assessments are sometimes controversial.

Since the implementation of The National Environmental Policy Act, environmental impact assessment (EIA) has become an increasingly important activity in project planning processes throughout the country. The increasing use of EIA has given rise to an enormous literature and one of the main component of the literature has been the publication of articles and reports describing and discussing methods of implementing comprehensive EIAs.

Methods and techniques are mostly concerned with the practical and technical activities involved in implementing EIA. Both methods and techniques are used in the assessment of specific projects whereas the EIA procedures apply to all projects within a particular national, regional or local-level jurisdiction.

Environmental Assessment (EA) is systemic reproducible and interdisciplinary evaluation of the potential effects of a proposed action and has practical alternatives on the physical, chemical, biological, cultural and socio-economic attributes of a particular geographical area, However, national policy is an important consideration in the effectiveness of an EA programme. EIA essentially involves four steps, viz. baseline data collection, impact identification, prediction and evaluation. The basic framework of the EIA for sustainable development has been adopted in series of Environmental settings, forecastings and assessing impact.

Describing the Environmental Setting

The description of the environmental setting sets forth in detail the characteristics of the area in which the proposed action could occur. This description should be of the study area which is defined area within which all effects, impacts, features and potential compensation efforts would occur from a proposed action and its alternatives. The level of detail of study area should be sufficient to convey to reader or reviewer the precise nature of natural and human resources potentially affected by the proposed action and alternatives. The study also provides base line data with which environmental impacts can be predicted, and against which the predicted impacts of the proposed action can be compared.

The approach commonly adopted in treating this aspect of EA is the subdivision of the environmental setting into a logical and hierarchial set of categories. The major categories would likely include the following:

Geology

Geological provinces, bedrock formations, history of geological stability or instability.

Topography

General topography of region, specific topography of region, specific topography of project area.

Soils

Soils' mapping, soil series properties, constraints to development.

Groundwater Resources

Nature of water-bearing formations, recharge rates, sustainable safe yields, locations and depths of existing wells, quality.

Surface Water Resources

Drainage basins and sub-basins, named and unnamed water bodies and water courses, regulatory classificaton of water bodies, flow regimes, water quality data and evaluation and identification of existing permitted discharges to surface waters.

Terrestrial Resources

Spatial arrangement of vegetative community types, vegetative species-abundance listings, wildlife species-abundance listings, records of endangered plant and animal species.

Aquatic Communities

Nature of aquatic habitats, species abundance listings for aquatic micro invertebrates and fish communities, ecological indexing of community data.

Environmentally Sensitive Areas

Identification of wetlands, floodplains, steep slopes, stands of mature vegetation, aquifer recharge areas, areas of high water table, areas of rock outcrop, prime agricultural lands and mines.

Air Quality

Regional quality and trends, data from local monitoring stations, reported exceedances of standards.

Land Use

Exciting patterns of land use in region, regional planning for future use, zoning.

Demography

Censused or estimated population, recent trends and projections for future populations.

Sound Levels

Existing sound levels, sources of sound.

Socio-Economics

Economic and social structure of communities, tax rates, characteristics of development.

Infrastructural Services

Nature and status of human services such as police and fire protection, hospitals, schools, utilities.

Transportation

Layout and function of roadways, railways, airports; existing and projeted capacities and demands.

Cultural Resources

Location and characterization of identified cultural resources (archaeological, historical, cultural landmark) potential for unidentified resources to be present in project area.

Project Economics

Comparative analysis of proposed alternatives with present worth cost effective criteria, cost/benefit criteria, or other methods.

Forecasting and Assessing Impacts

Biological Environment Including Terrestrial and Aquatic Habitats- Assessment of these natural focuses initially on the nature and distribution of existing habitats and biotic communities and the selective forces related directly or indirectly to the proposed action should be considered. If the effects relate to the proposed action are likely to induce a "change of state" in one or more of the affected habitats or biotic communities, the impact could be considered significant. A change in state could be anticipated by forecasting with ecological indices or habitat evaluation models: for example, if the project could increase phosphorus levels in a lake, the post development tropic state of the lake can be predicted. Likewise, if one or more habitats are likely to be reduced significantly in size or habitat value, such an effect might be deemed significant. The use of habitat evaluation models can summarize present and future habitat values and can quantify the degree of change likely to occur, if the action were implemented.

Water Environment

Assessment of water resources focuses on the identification of surface water and groundwater resources, their existing quality and use, and the regulatory standards applied to them. If water resources are to be used, the capacity of the resource to accommodate this additional use should be evaluated. Where water might be drawn from subsurface aquifers and then, after use, discharged to surface water, the effects of this "diversion" of water from existing hydrological cycle should be examined. The quality of water might be altered by the proposed action from a waste water discharge. Water quality models can be run to forecast impacts, and then future water quality should be compared to regulatory standards and to the tolerances of organisms using the water bodies as habitat.

Air Environment

Assessment of the air environment begins with the documentation of existing air quality, desirable air quality and prevailing regulatory standards. The potential effect of the proposed action should be understood well enough to model the concentrations of important air-borne contaminants that would be expected if the action were implemented. These anticipated concentrations can be compared to regulatory standards and to standards or guidelines for safe human exposure. Because the effect of air quality changes are generally gauged in terms of the affected human population, the determination of significance in impacts may be more clearly discernible by human health standards.

Noise Environment

Assessment of the noise environment is similar to that of air; the existing sound levels can be compared to regulatory standards and the increases in sound levels, if any, can be gauged for significance against these regulatory standards. As with air assessments, sound level assessments are generally considered in the perspective of the human population affected.

Socio-economic Environment

Assessment of socio-economic impacts should consider existing demography, land values, income distribution, tax schedules and other related information about the structure and function of the human communities affected by proposed action. The changes in these properties resulting from implementation of the action can often be estimated as monetary costs of benefits, resulting in a net gain or loss of socio-economic assets.

Cultural Environment

Assessment of cultural impacts focuses on the existence of recorded archaeological, historic, or cultural resources have been formally described, literature and field studies may be necessary to describe sufficiently the nature and extent of such resources. The assessment of impacts should consider the importance of cultural resources destroyed by the action and the feasibility of recovering or conserving all or portions of resources.

Significance of Effect on the Environment

There are particular type of actions that, though of somewhat more limited scope than in the sense as discussed above, have the potential to exert significant effects on the environment. These potential effects may be related to high sensitivity of the resources of disturbance, to the specific nature of the impacts, to the potential secondary and cumulative impacts. The principal concern in these types of actions is that the environmental effects could cause irrevocable losses of important natural resources or irreversible effects on human health.

Highly sensitive environmental resources may be certain natural environments such as major wetland complexes (either inland or coastal), the flood plains of major rivers, large tracts of prime agricultural land or forest land, habitats with threatened species or regions with extensive and important cultural resources. In addition, highly sensitive resources may be human resources; the potential risk of an action may be a function of the number of "sensitive receptors" in the affected area (sensitive receptors here being such things as schools, potable water supplies, recreational areas and housing for elderly). Increased levels of air pollution, for example, are likely to be more significant if the young, the old, or the ill- the more sensitive human receptors- will be exposed.

The nature of the potential effects will affect the EA. Discharges of natural wastes (e.g., domestic sewage) will produce undesirable changes in surface or groundwater characteristics, but these changes are qualitatively different from those potentially resulting from the release of a synthetic chemical that is both hazardous and persistent. The effects of the discharge of the natural wastes are reversible, atleast in part, while contamination of surface or groundwaters with hazardous and persistent chemical compounds present a significant problem over a longer-term. The significance of such effect should also be considered in terms of time period. As the length of the exposure or the frequency of exposure increases, so will the likelihood of adverse impacts.

Characteristics of EIA

The EIA gives a clear-cut picture of important environmental considerations which are to be incorporated in the decision-making process while setting up of the major projects.

It has many important characteristics:

1. Provision of systemic evaluation of almost all significant environmental consequences of a developmental project.
2. EIA is a structured, systemic and comprehensive approach.
3. All EIA processes draw driving force by legislation.
4. EIA is one of the most valuable and interdisciplinary decision-making tools.
5. EIA can force the policy makers to reconsider the policy proposals.
6. It can be used to ensure regional planning for substantial development.
7. It ensures the accountability of decision-makers to public.
8. It provides possible alternate development options against identified environmental impacts.
9. EIA document should be circulated for objective review of its results by others.
10. EIA cannot be taken in isolation, rather it is related with other factors, viz., Environmental Statement, Environmental Audit and New System of National Accounts.
11. It seeks public participation in decision-making.

Objectives of EIA

There are three primary objectives of EIA

(i) EIA predicts the likely environmental impacts of projects.
(ii) It presents these predictions and options to decision makers.
(iii) It finds ways to reduce unacceptable impacts to shape the project so that it suits the local environment.

AROUND THE WORLD

At the federal level, EIA provisions are contained within the EPBC Act (the *Environment Protection and Biodiversity Conservation Act, 1999*). At the State [i.e. provincial] level, each jurisdiction has EIA provisions typically contained in land use planning law. For example, in New South Wales (NSW), EIA is performed under either Part 3A, Part 4 or Part 5 of the *Environmental Planning and Assessment Act 1979*, depending on the type of development. Overlap between federal and state requirements is addressed via bilateral agreements or one off accreditation of state processes, as provided for in the EPBC Act. The Canadian Environmental Assessment Act (CEAA) is the legal basis for the federal environmental assessment (EA) process. CEAA came into force in 1995. Legislative amendments were introduced in 2001 and came into force on October 30, 2003. EA is defined as a planning tool to identify, understand, assess and mitigate, where possible, the environmental effects of a project. Under the CEAA, Natural Resources Canada (NRCan) is required to undertake an EA for projects relating to a physical work and for any proposed physical activity listed in the Inclusion List Regulations where it exercises one or more of the following CEAA triggers:

- Proposes or undertakes a project.
- Grants money or any other form of financial assistance to a project.
- Grants an interest in the land to enable a project to be carried out.
- Exercises a regulatory duty in relation to a project, such as issuing a permit or license that is included in the Law List Regulations.

If NRCan exercises one or more of the above-mentioned triggers, it becomes a Responsible Authority (RA) under CEAA. As an RA, NRCan must ensure that an EA is carried out in accordance with the CEAA and must consider the EA findings before a decision is made that will allow the project to proceed.

The Environmental Impact Assessment Law (EIA Law) requires an environmental impact assessment to be completed prior to project construction. However, if a developer completely ignores this requirement and builds a project without submitting an environmental impact statement, the only penalty is that the Environmental Protection Bureau (EPB) may require the developer to do a make-up environmental assessment. If the developer does not complete this make-up assessment within the designated time, only then is the EPB authorized to fine the developer.

A joint investigation by SEPA and the Ministry of Land and Resources in 2004 showed that 30-40% of the mining construction projects went through the procedure of environment impact assessment as required, while in some areas only 6-7% did so. This partly explains why China has witnessed so many mining accidents in recent years.

Progresses were made in the Environmental protection issue during the 8th Five Year Plan (1992-1997). The following developments in Environmental protection were achieved during that time:

- Formulation of Environmental Protection Act, 1992.
- Establishment of Ministry of Population and Environment.
- Development of National Environmental Policies and Action Plan, EIA guidelines developed.
- Consideration of environmental concerns in hydropower projects.
- Development of industrial, irrigation and agricultural policies that undertook environmental concerns.

Chapter 19

Soil Pollution

Soil is one of the most significant ecological factors which is derived from the transformation of surface rocks. The word *soil* is derived from the Latin word solum, which means earthly materials in which plants grow. The study of soil is commonly referred to as soil science or Pedology (pedos=earth) or Edaphology (edaphos = soil + logos = study). Soil can be defined as weathered organisms and their products of decay intermixed.

Soil is nothing but a substratum on which plants depend on their nutrients, waters and mineral supply and anchorage. It constitutes an important medium wherein numerous animals live. In fact, the soil of a nation is the most valuable material heritage.

Definition of Soil Pollution

Soil pollution comprises the pollution of soils with materials, mostly chemicals that are out of place and are present at concentrations higher than normal which may have adverse effects on human and other organisms. There are different opinions on how to characterize a pollutant, while some consider the use of pesticide acceptable if their effect does not exceed the intended result, other do not consider the use of pesticides or even chemical fertilizers acceptable. However, soil pollution is also caused by means other than the direct addition of xenobiotic (man-made) chemicals such as agricultural runoff waters, industrial waste materials, acidic precipitates and radioactive fallouts.

Both organic (those that contain carbon) and inorganic (those that don't) contaminants are important in soil. The most prominent chemical groups of organic contaminants are fuel hydrocarbons, polynuclear aromatic hydrocarbons (PAHs), polychlorinated biphenyl (PCBs), chlorinated aromatic compounds, detergents and pesticides. Inorganic species include nitrates, etc.

Chapter 20

Air Pollution and Noise

Preamble

To systematically address the air quality impacts of potential projects or activities, it is necessary to be familiar with basic information regarding air pollution. Accordingly, this section includes a definition of air pollution and information on the types and effects of specific air pollutants, along with summary information on air pollution sources. "Air Pollution" can be defined as the presence in the outdoor atmosphere of one or more contaminants in such quantities and of such duration as may be injurious to human, plant, or animal life, or to materials, or which may unreasonably interfere with the comfortable enjoyment of life or property, or the conduct of business.

Sources of Air Pollutants

Air pollutant sources can be categorized from several perspectives, including the type of source, their frequency of occurrence and spatial distribution, and the types of emissions. Characterization by source type can be delineated as arising from natural sources or from man-made sources. 'Natural Sources' include plant pollens, windblown dust, volcanic eruptions and lightning generated forest fires. 'Man-made source' can include transportation vehicles, industrial processes, power plants, construction activities and military training activities.

Source characterization according to number and spatial distribution can include such categories as single or point sources (stationary), area or multiple sources (stationary or mobile) and line sources. 'Point sources' are characteristics of pollutant emissions from industrial process stacks, as well as fuel-combustion facility stacks. 'Area sources' include vehicular traffic, fugitive-dust emissions from resource material stockpiles or construction, or military training activities over large geographical areas.

Effects of Air Pollutants

Air pollution effects may also be divided into several categories, with such effects encompassing those that are health-related as well as those associated with damage to materials or which cause decrease in atmospheric aesthetic features. Examples of effects on human health include eye irritation, headaches and aggravation of respiratory difficulties. Plants and crops have been subjected to undesirable consequences of air pollution, including abnormal growth patterns, leaf discoloration or spotting and death. Property damage include property devaluation because of odors, deterioration of materials such as concrete statutory, discoloration of painted surfaces. The aesthetic effects include reductions in visibility, discoloration of air, photochemical smog- related traffic disruptions at airports and the general nuisance aspects of odors and duct.

Identification of the Types and Quantities of Air Pollutants and their Impacts

An appropriate initial step when analyzing any proposed project activity is to consider what types of air pollutants might be emitted during construction or operational phases of the proposed project activity, and the quantities in which such air pollutants are expected to occur. Use of emission-factor information organized according to project type or activity is a suggested approach. An 'emission factor' is the average rate at which a pollutant is released into the atmosphere as a result of some activity, such as combustion or industrial production, divided by the level of that activity (U.S.EPA, 1973). "Emission factors" relate the type and quantities of pollutants emitted to indicators such as production capacity, quantity of fuel burned, or vehicle miles traveled by an automobile.

Description of Existing Air Quality Conditions

Existing air quality conditions can be described in terms of ambient air quality data, emission inventories and meteorological information, which relates to atmospheric dispersion. For a quick assessment and prediction of any air quality survey study data on immediate atmospheric environment, meteorological details such as temperature, humidity, wind speed and directions are very essential, since they play an important role in transport, dilution and dispersion of air pollutants.

Frequency and Method of Air Quality Measurements

The present study of air quality measurement in terms of air pollution concentration has been obtained by continuous monitoring for a period of eight hours at all directions and at the proposed project site. The annual mean wind direction pattern has been compiled from the long-term data made available from meteorological department. The sampling directions around the project site has been selected so as to reflect the impact of anthropogenic activities such as emissions from transportation, generation of dust with movement of vehicles, emissions from industrial and domestic activities. Considering the facilities and significant impact on air quality, we have selected Suspended Particulate Matter (SPM), Sulphur dioxide (SO_2), Nitrogen Oxides (Nox) and Carbon monoxide (CO).

METEOROLOGY

Effect of pollution on receptors, animate and inanimate depends on atmospheric conditions. Therefore knowledge of meteorological characteristics in a study area is important. Wind speed and wind direction are called primary meteorological parameters because the dispersion and dilution of the pollutants depends mainly on these.

Wind speed determines the rate of dilution and travel time of air pollutant. Travel time is the time required by the pollutant to reach a given receptor. Wind direction is important in air pollution problems since it determines the direction of transport of air pollutants. The mean wind direction over a period of time will identify the receptors which will be affected by a particular source during that time period. In other words, the mean wind direction will be indicative of the direction of travel of the pollutant. Wind rose are the diagramatic representations of wind speed and wind direction for quick understanding. The windrose are drawn on the basis of three hourly data collected by the meteorological office. Ambient air quality monitoring refers to appropriate sampling and analysis to establish the ambient concentrations of specific pollutants. Targeted monitoring might be desirable in order to verify the experienced changes in air quality concentrations for those pollutants determined to be of concern. The primary sources of information on air quality standards, criteria and policies will be the relevant local, state and federal agencies which have a mandate for overseeing the air resources of the geographical area. Documentation of this information will allow the determination of the significance of air quality impacts incurred during projects or activities and will aid in deciding between alternative actions or in assessing the need for mitigating measures for a given alternative.

Air quality impact prediction can be based on several approaches including mass-balances, the use of mathematical models and other considerations. Air pollutant emissions from the construction or operational phase of a project activity can be considered in relation to the existing emission inventory for the study area. Particular attention should be given to releases of air toxics and the resultant comparison of maximum ground level concentrations to appropriate ambient air quality guidelines or standards. In conjunction with these analyses, it is important to identify prevention, control and mitigation measures for releases of air toxics.

NOISE

Noise can be defined as unwanted sound or sound in the wrong place at wrong time. It can also be defined as any sound that is undesirable because it interferes with speech and hearing, is intense enough to damage hearing. The definition of noise as unwanted sound implies that it has an adverse effect on human beings and their environment including land structures and domestic animals. Noise can also disturb natural wildlife and ecological systems. The number of compressions and rarefactions of the air molecules in a unit of time is described as its frequency and is expressed in hertz. Humans can detect sounds with frequencies ranging about 16 to 20,000 Hz. In most noise considerations, the "A-weighted-sound level" scale is used and the units are weighted decibels (dBA). The weighted sound level or noise level meters have an A-weighting network, thus yielding A-weighted dB or dBA readings.

To provide a basis for addressing noise environment impacts, the following methodology is suggested for the planning and conduction of impact studies:

1. The first step is to determine the potential impacts of the proposed project on the noise environment, which requires the identification of the noise levels associated with the project.
2. In analyzing the potential impacts of a project, it is necessary to consider the study area or region of influence associated with the noise emissions. The delineation of a study area can be based upon the boundaries of the land associated with the project or by also considering the area of noise influence within the vicinity of the project.
3. Predicting the propagation of noise from a source and determining the type of affected land uses with the help of noise standards/policies by the state and federal agencies which have a mandate overseeing on the noise environment of the study area.
4. The final step is the assessment impact significance with the identification and incorporation of mitigation measures. Assessment refers to the interpretation of the significance of anticipated changes related to the proposed project.

The important sources of noise presently in the study area are transportation, industry like sugar, detergent, cement etc.

Chapter 21

Water Pollution

Water pollution is the contamination of water bodies such as lakes, rivers, oceans, and groundwater caused by human activities, which can be harmful to organisms and plants that live in these water bodies. It occurs when pollutants are discharged directly into water bodies without treating it first.

Water pollution is a major problem in the global context. It has been suggested that it is the leading worldwide cause of deaths and diseases, and that it accounts for the deaths of more than 14,000 people daily. In addition to the acute problems of water pollution in developing countries, industrialized countries continue to struggle with pollution problems as well. Water is typically referred to as polluted when it is impaired by anthropogenic contaminants and either does not support a human use, like serving as drinking water, and/or undergoes a marked shift in its ability to support its constituent biotic communities, such as fish. Natural phenomena such as volcanoes, algae blooms, storms, and earthquakes also cause major changes in water quality and the ecological status of water. Water pollution has many causes and characteristics.

Water Pollution Categories

Surface water and groundwater have often been studied and managed as separate resources, although they are interrelated. Sources of surface water pollution are generally grouped into two categories based on their origin.

Point Source Pollution

Point source pollution refers to contaminants that enter a waterway through a discrete conveyance, such as a pipe or ditch. Examples of sources in this category include discharges from a sewage treatment plant, a factory, or a city storm drain.

Non-point Source Pollution

Non-point source (NPS) pollution refers to diffuse contamination that does not originate from a single discrete source. NPS pollution is often a cumulative effect of small amounts of contaminants gathered from a large area. Nutrient runoff in storm water from "sheet flow" over an agricultural field or a forest are sometimes cited as examples of NPS pollution.

Contaminated storm water washed off from parking lots, roads and highways, called urban runoff, is sometimes included under the category of NPS pollution. However, this runoff is typically channeled into storm drain systems and discharged through pipes to local surface waters, and is a point source.

Groundwater Pollution

Interactions between groundwater and surface water are complex. Consequently, groundwater pollution, sometimes referred to as groundwater contamination, is not as easily classified as surface water pollution. By its very nature, groundwater aquifers are susceptible to contamination from sources that may not directly affect surface water bodies, and the distinction of point vs. non-point source may be irrelevant. A spill of a chemical contaminant on soil, located away from a surface water body, may not necessarily create point source or non-point source pollution, but nonetheless may contaminate the aquifer below. Analysis of groundwater contamination may focus on soil characteristics and hydrology, as well as the nature of the contaminant itself.

Causes of Water Pollution

The specific contaminants leading to pollution in water include a wide spectrum of chemicals, pathogens, and physical or sensory changes such as elevated temperature and discoloration. While many of the chemicals and substances that are regulated may be naturally occurring (calcium, sodium, iron, manganese, etc.), the concentration is often the key in determining what is a natural component of water, and what is a contaminant. Oxygen-depleting substances may be natural materials, such as plant matter (e.g. leaves and grass) as well as man-made chemicals. Other natural and anthropogenic substances may cause turbidity (cloudiness) which blocks light and disrupts plant growth, and clogs the gills of some fish species. Many of the chemical substances are toxic. Pathogens can produce water-borne diseases in either human or animal hosts. Alteration of water's physical chemistry include acidity (change in pH), electrical conductivity, temperature, and eutrophication. Eutrophication is the fertilization of surface water by nutrients that were previously scarce.

Pathogens

Coliform bacteria are a commonly used bacterial indicator of water pollution, although not an actual cause of disease. Other microorganisms sometimes found in surface waters which have caused human health problems include:

- *Cryptosporidium parvum*
- *Giardia lamblia*
- *Salmonella*
- *Novovirus* and other viruses
- Parasitic worms (helminths)

High levels of pathogens may result from inadequately treated sewage discharges. This can be caused by a sewage plant designed with less than secondary treatment (more typical in less-developed countries). In developed countries, older cities with aging infrastructure may have leaky sewage collection systems (pipes, pumps, valves), which can cause sanitary sewer overflows. Some cities also have combined sewers, which may discharge untreated sewage during rain storms.

Pathogen discharges may also be caused by poorly-managed livestock operations.

Chemical and Other Contaminants

Contaminants may include organic and inorganic substances.

Organic water pollutants include:

- Detergents
- Disinfection of by-products found in chemically disinfected drinking water, such as chloroform
- Food processing waste, which can include oxygen-demanding substances, fats and grease
- Insecticides and herbicides, a huge range of organohalides and other chemical compounds
- Petroleum hydrocarbons, including fuels (gasoline, diesel fuel, jet fuels, and fuel oil) and lubricants (motor oil), and fuel combustion by products, from storm water runoff
- Tree and brush debris from logging operations
- Volatile organic compounds (VOCs), such as industrial solvents, from improper storage. Chlorinated solvents, which are dense non-aqueous phase liquids (DNAPLs), may fall to the bottom of reservoirs, since they don't mix well with water and are denser.
- Various chemical compounds found in personal hygiene and cosmetic products

Inorganic water pollutants include:

- Acidity caused by industrial discharges (especially sulfur dioxide from power plants)
- Ammonia from food processing waste
- Chemical waste as industrial by-products
- Fertilizers containing nutrients—nitrates and phosphates—which are found in storm water runoff from agriculture, as well as commercial and residential use
- Heavy metals from motor vehicles (via urban storm water runoff) and acid mine drainage
- Silt (sediment) in runoff from construction sites, logging, slash and burn practices or land clearing sites

Macroscopic pollution—large visible items polluting the water—may be termed "floatables" in an urban storm water context, or marine debris when found on the open seas, and can include such items as:

- Trash (e.g. paper, plastic, or food waste) discarded by people on the ground, and that are washed by rainfall into storm drains and eventually discharged into surface waters
- Nurdles, small ubiquitous water borne plastic pellets
- Shipwrecks, large derelict ships

Thermal Pollution

Thermal pollution is the rise or fall in the temperature of a natural body of water caused by human influence. A common cause of thermal pollution is the use of water as a coolant by power plants and industrial manufacturers. Elevated water temperatures decreases oxygen levels (which can kill fish) and affects ecosystem composition, such as invasion by new thermophilic species. Urban runoff may also elevate temperature in surface waters.

Thermal pollution can also be caused by the release of very cold water from the base of reservoirs into warmer rivers.

Transport and Chemical Reactions of Water Pollutants

Most water pollutants are eventually carried by rivers into the oceans. In some areas of the world, the influence can be traced hundred miles from the mouth by studies using hydrology transport models. Advanced computer models such as SWMM or the DSSAM Model have been used in many locations worldwide to examine the fate of pollutants in aquatic systems. Indicator filter feeding species such as copepods have also been used to study pollutant fates in the New York Bight, for example. The highest toxin loads are not directly at the mouth of the Hudson River, but 100 kilometers south, since several days are required for incorporation into planktonic tissue. The Hudson discharge flows south along the coast due to coriolis force. Further southern are areas of oxygen depletion, caused by chemicals using up oxygen and by algae blooms, caused by excess nutrients from algal cell death and decomposition. Fish and shell fish kills have been reported, because toxins climb the food chain after small fish consume copepods, then large fish eat smaller fish, etc. Each successive step up the food chain causes a stepwise concentration of pollutants such as heavy metals (e.g. mercury) and persistent organic pollutants such as DDT. This is known as biomagnification, which is occasionally used interchangeably with bioaccumulation.

Large gyres (vortexes) in the oceans trap floating plastic debris. The North Pacific Gyre, for example, has collected the so-called "Great Pacific Garbage Patch" that is now estimated at 100 times the size of Texas. Many of these long-lasting pieces wind up in the stomachs of marine birds and animals. This results in obstruction of digestive pathways which leads to reduced appetite or even starvation.

Many chemicals undergo reactive decay or chemically change especially over long periods of time in groundwater reservoirs. A noteworthy class of such chemicals is the chlorinated hydrocarbons such as trichloroethylene (used in industrial metal degreasing and electronics manufacturing) and tetrachloroethylene used in the dry cleaning industry. Both of these chemicals, which are carcinogens themselves, undergo partial decomposition reactions, leading to new hazardous chemicals (including dichloroethylene and vinyl chloride).

Groundwater pollution is much more difficult to abate than surface pollution because groundwater can move great distances through unseen aquifers. Non-porous aquifers such as clays partially purify water of bacteria by simple filtration (adsorption and absorption), dilution, and, in some cases, chemical reactions and biological activity: however, in some cases, the pollutants merely transform to soil contaminants. Groundwater that moves through cracks and caverns is not filtered and can be transported as easily as surface water. In fact, this can be aggravated by the human tendency to use natural sinkholes as dumps in areas of Karst topography.

There are a variety of secondary effects stemming not from the original pollutant, but a derivative condition. An example is silt-bearing surface runoff, which can inhibit the penetration of sunlight through the water column, hampering photosynthesis in aquatic plants.

Measurement of Water Pollution

Water pollution may be analyzed through several broad categories of methods: physical, chemical and biological. Most methods involve collection of samples, followed by specialized analytical tests. Some methods may be conducted *in situ*, without sampling, such as temperature. Government agencies and research organizations have published standardized, validated analytical test methods to facilitate the comparability of results from disparate testing events.

Sampling

Sampling of water for physical or chemical testing can be done by several methods, depending on the accuracy needed and the characteristics of the contaminant. Many contamination events are sharply restricted in time, most commonly in association with rain events. For this reason "grab" samples are often inadequate for fully quantifying contaminant levels. Scientists gathering this type of data often employ auto-sampler devices that pump increments of water at either time or discharge intervals.

Sampling for biological testing involves collection of plants and/or animals from the surface water body. Depending on the type of assessment, the organisms may be identified for biosurveys (population counts) and returned to the water body, or they may be dissected for bioassays to determine toxicity.

Physical Testing

Common physical tests of water include temperature, solids concentration and turbidity.

Chemical Testing

Water samples may be examined using the principles of analytical chemistry. Many published test methods are available for both organic and inorganic compounds. Frequently-used methods include pH, biochemical oxygen demand (BOD), chemical oxygen demand (COD), nutrients (nitrate and phosphorus compounds), metals (including copper, zinc, cadmium, lead and mercury), oil and grease, total petroleum hydrocarbons (TPH), and pesticides.

Biological Testing

Biological testing involves the use of plant, animal, and/or microbial indicators to monitor the health of an aquatic ecosystem.

Control of Water Pollution

Domestic Sewage

In urban areas, domestic sewage is typically treated by centralized sewage treatment plants. Municipal treatment plants are designed to control conventional pollutants: BOD and suspended solids. Well-designed and operated systems (i.e., secondary treatment or better) can remove 90 percent or more of these pollutants. Some plants have additional sub-systems to treat nutrients and pathogens. Most municipal plants are not designed to treat toxic pollutants found in industrial wastewater.

Cities with sanitary sewer overflows or combined sewer overflows employ one or more engineering approaches to reduce discharges of untreated sewage, including:

- utilizing a green infrastructure approach to improve stormwater management capacity throughout the system
- repair and replacement of leaking and malfunctioning equipment
- increasing overall hydraulic capacity of the sewage collection system (often a very expensive option).

A household or business not served by a municipal treatment plant may have an individual septic tank, which treats the wastewater on site and discharges into the soil. Alternatively, domestic wastewater may be sent to a nearby privately-owned treatment system (e.g. in a rural community).

Industrial Wastewater

Some industrial facilities generate ordinary domestic sewage that can be treated by municipal facilities. Industries that generate wastewater with high concentrations of conventional pollutants (e.g. oil and grease), toxic pollutants (e.g. heavy metals, volatile organic compounds) or other non-conventional pollutants such as ammonia, need specialized treatment systems. Some of these facilities can install a pre-treatment system to remove the toxic components, and then send the partially-treated wastewater to the municipal system. Industries generating large volumes of wastewater typically operate their own complete on-site treatment systems.

Some industries have been successful at redesigning their manufacturing processes to reduce or eliminate pollutants, through a process called pollution prevention.

Heated water generated by power plants or manufacturing plants may be controlled with:

- cooling ponds, man-made bodies of water designed for cooling by evaporation, convection, and radiation
- cooling towers, which transfer waste heat to the atmosphere through evaporation and/or heat transfer
- cogeneration, a process where waste heat is recycled for domestic and/or industrial heating purposes.

Agricultural Wastewater

Nonpoint Source Controls

Sediment (loose soil) washed off fields is the largest source of agricultural pollution. Farmers may utilize erosion controls to reduce runoff flows and retain soil on their fields. Common techniques include contour plowing, crop mulching, crop rotation, planting perennial crops and installing riparian buffers.

Nutrients (nitrogen and phosphorus) are typically applied to farmland as commercial fertilizer; animal manure; or spraying of municipal or industrial wastewater (effluent) or sludge. Nutrients may also enter runoff from crop residues, irrigation water, wildlife, and atmospheric deposition. Farmers can develop and implement nutrient management plans to reduce excess application of nutrients.

To minimize pesticide impacts, farmers may use Integrated Pest Management (IPM) techniques (which can include biological pest control) to maintain control over pests, reduce reliance on chemical pesticides, and protect water quality.

Point Source Wastewater Treatment

Farms with large livestock and poultry operations, such as factory farms, are called concentrated animal feeding operations or confined animal feeding operations in the U.S. and are being subject to increasing government regulation. Animal slurries are usually treated by containment in lagoons before disposal by spray or trickle application to grassland. Constructed wetlands are sometimes used to facilitate treatment of animal wastes, as are anaerobic lagoons. Some animal slurries are treated by mixing with straw and composted at high temperature to produce a bacteriologically sterile and friable manure for soil improvement.

Construction Site Stormwater

Sediment from construction sites is managed by installation of:

- erosion controls, such as mulching and hydroseeding, and
- sediment controls, such as sediment basins and silt fences.

Discharge of toxic chemicals such as motor fuels and concrete washout is prevented by use of:

- spill prevention and control plans, and
- specially-designed containers (e.g. for concrete washout) and structures such as overflow controls and diversion berms.

Urban Runoff (Stormwater)

Effective control of urban runoff involves reducing the velocity and flow of stormwater, as well as reducing pollutant discharges. Local governments use a variety of stormwater management techniques to reduce the effects of urban runoff. These techniques are called best management practices (BMPs), may focus on water quantity control, while others focus on improving water quality, and some perform both functions.

Pollution prevention practices include low impact development techniques, installation of green roofs and improved chemical handling (e.g., management of motor fuels and oil, fertilizers and pesticides). Runoff mitigation systems include infiltration basins, bioretention systems, constructed wetlands, retention basins and similar devices. Thermal pollution from runoff can be controlled by stormwater management facilities that absorb the runoff or direct it into groundwater, such as bioretention systems and infiltration basins. Retention basins tend to be less effective at reducing temperature, as the water may be heated by the sun before being discharged to a receiving stream.

Regulatory Framework

In developed countries, the primary focus of legislation and efforts to curb water pollution for the past several decades was first aimed at point sources. As many point sources have been effectively regulated—principally factories and sewage treatment plants—greater attention has been placed on controlling municipal and industrial stormwater discharges, and NPS contributions.

Marine Pollution

Marine pollution occurs when harmful effects, or potentially harmful effects, can result from the entry into the ocean of chemicals, particles, industrial, agricultural and residential waste, or the spread of invasive organisms. Most sources of marine pollution are land based. The pollution often comes from non-point sources such as agricultural runoff and wind blown debris. Many potentially toxic chemicals adhere to tiny particles which are then taken up by plankton and benthos animals, most of which are either deposit or filter feeders. In this way, the toxins are concentrated upward within ocean food chains. Many particles combine chemically in a manner highly depletive of oxygen, causing estuaries to become anoxic. When pesticides are incorporated into the marine ecosystem, they quickly become absorbed into marine food webs. Once in the food webs, these pesticides can cause mutations, as well as diseases, which can be harmful to humans as well as the entire food web.

Toxic metals can also be introduced into marine food webs. These can cause a change to tissue matter, biochemistry, behaviour, reproduction, and suppress growth in marine life. Also, many animal feeds have a high fish meal or fish hydrolysate content. In this way, marine toxins can be transferred to land animals, and appear later in meat and dairy products.

Pathways of Pollution

Bjorn Jenssen (2003) notes in his article, "Anthropogenic pollution may reduce biodiversity and productivity of marine ecosystems, resulting in reduction and depletion of human marine food resources". There are many different ways to categorize, and examine the inputs of pollution into our marine ecosystems. Patin notes that generally there are three main types of inputs of pollution into the ocean: direct discharge of waste into the oceans, runoff into the waters due to rain, and pollutants that are released from the atmosphere.

One common path of entry by contaminants to the sea are rivers. Pollution is often classed as point source or non-point source pollution. Point source pollution occurs when there is a single, identifiable, and localized source of the pollution. An example is directly discharging sewage and industrial waste into the ocean. Pollution such as this occurs particularly in developing nations. Non-point source pollution occurs when the pollution comes from ill-defined and diffuse sources. These can be difficult to regulate. Agricultural runoff and wind blown debris are prime examples.

Pollution from Ships

Ships can pollute waterways and oceans in many ways. Oil spills can have devastating effects. While being toxic to marine life, polycyclic aromatic hydrocarbons (PAHs), the components in crude oil, are very difficult to clean up, and last for years in the sediment and marine environment.

Discharge of cargo residues from bulk carriers can pollute ports, waterways and oceans. In many instances, vessels intentionally discharge illegal wastes despite foreign and domestic regulation prohibiting such actions. Ships create noise pollution that disturbs natural wildlife, and water from ballast tanks can spread harmful algae and other invasive species.

Meinesz believes that one of the worst cases of a single invasive species causing harm to an ecosystem can be attributed to a seemingly harmless jellyfish. *Mnemiopsis leidyi*, a species of comb jellyfish that spread so it now inhabits estuaries in many parts of the world. It was first introduced in 1982, and thought to have been transported to the Black Sea in a ship's ballast water. The population of the jellyfish shot up exponentially and, by 1988, it was wreaking havoc upon the local fishing industry. "The anchovy catch fell from 204,000 tonnes in 1984 to 200 tons in 1993; sprat from 24,600 tonnes in 1984 to 12,000 tonnes in 1993; horse mackerel from 4,000 tonnes in 1984 to zero in 1993." Now that the jellyfish have exhausted the zooplankton, including fish larvae, their numbers have fallen dramatically, yet they continue to maintain a stranglehold on the ecosystem. Invasive species can take over once occupied areas, facilitate the spread of new diseases, introduce new genetic material, alter underwater seascapes and jeopardize the ability of native species to obtain food.

Plastic Debris

Marine debris is mainly discarded human rubbish which floats on, or is suspended in the ocean. Eighty percent of marine debris is plastic - a component that has been rapidly accumulating since the end of World War II. The mass of plastic in the oceans may be as high as one hundred million metric tons. Discarded plastic bags, six-pack rings and other forms of plastic waste which finish up in the ocean present dangers to wildlife and fisheries. Aquatic life can be threatened through entanglement, suffocation, and ingestion. Fishing nets, usually made of plastic, can be left or lost in the ocean by fishermen. Known as ghost nets, these entangle fish, dolphins, sea turtles, sharks, dugongs, crocodiles, seabirds, crabs, and other creatures, restricting movement, causing starvation, laceration and infection, and, in those that need to return to the surface to breathe, suffocation.

Many animals that live on or in the sea consume flotsam by mistake, as it often looks similar to their natural prey. Plastic debris, when bulky or tangled, is difficult to pass, and may become permanently lodged in the digestive tracts of these animals, blocking the passage of food and causing death through starvation or infection. Plastics accumulate because they don't biodegrade in the way many other substances do. They will photodegrade on exposure to the sun, but they do so properly only under dry conditions, and water inhibits this process. In marine environments, photodegraded plastic disintegrates into ever smaller pieces while remaining polymers, even down to the molecular level. When floating plastic particles photodegrade down to zooplankton sizes, jellyfish attempt to consume them, and in this way the plastic enters the ocean food chain. Many of these long-lasting pieces end up in the stomachs of marine birds and animals, including sea turtles, and black-footed albatross.

Plastic debris tends to accumulate at the centre of ocean gyres. In particular, the Great Pacific Garbage Patch has a very high level of plastic particulate suspended in the upper water column. In samples taken in 1999, the mass of plastic exceeded that of zooplankton (the dominant animal life in the area) by a factor of six. Toxic additives used in the manufacture of plastic materials can leach out into their surroundings when exposed to water. Water-borne hydrophobic pollutants collect and magnify on the surface of plastic debris, thus making plastic far more deadly in the ocean than it would be on land. Hydrophobic contaminants are also known to bioaccumulate in fatty tissues, biomagnifying up the food chain and putting pressure on apex predators. Some plastic additives are known to disrupt the endocrine system when consumed, others can suppress the immune system or decrease reproductive rates. Floating debris can also absorb persistent organic pollutants from seawater, including PCBs, DDT and PAHs. Aside from toxic effects, when ingested some of these are mistaken by the animal brain for estradiol, causing hormone disruption in the affected wildlife.

Toxins

Apart from plastics, there are particular problems with other toxins that do not disintegrate rapidly in the marine environment. Examples of persistent toxins are PCBs, DDT, pesticides, furans, dioxins and phenols. Heavy metals are metallic chemical elements that have a relatively high density and are toxic or poisonous at low concentrations. Examples are mercury, lead, nickel, arsenic and cadmium. Such toxins can accumulate in the tissues of many species of aquatic life in a process called bioaccumulation. They are also known to accumulate in benthic environments, such as estuaries and bay muds: a geological record of human activities of the last century.

Eutrophication

Eutrophication is an increase in chemical nutrients, typically compounds containing nitrogen or phosphorus, in an ecosystem. It can result in an increase in the ecosystem's primary productivity (excessive plant growth and decay), and further effects are including lack of oxygen and severe reductions in water quality, fish, and other animal populations.

The biggest culprit are rivers that empty into the ocean, and with it the many chemicals used as fertilizers in agriculture as well as waste from livestock and humans. An excess of oxygen depleting chemicals in the water can lead to hypoxia and the creation of a dead zone.

Estuaries tend to be naturally eutrophic because land-derived nutrients are concentrated where runoff enters the marine environment in a confined channel. The World Resources Institute has identified 375 hypoxic coastal zones around the world, concentrated in coastal areas in Western Europe, the Eastern and Southern coasts of the US, and East Asia, particularly in Japan. In the ocean, there are frequent red tide algae blooms that kill fish and marine mammals and cause respiratory problems in humans and some domestic animals when the blooms reach close to shore.

In addition to land runoff, atmospheric anthropogenic fixed nitrogen can enter the open ocean. A study in 2008 found that this could account for around one third of the ocean's external (non-recycled) nitrogen supply and up to three percent of the annual new marine biological production. It has been been suggested that accumulating reactive nitrogen in the environment may have consequences as serious as putting carbon dioxide in the atmosphere.

Acidification

The oceans are normally a natural carbon sink, absorbing carbon dioxide from the atmosphere. Because the levels of atmospheric carbon dioxide are increasing, the oceans are becoming more acidic. The potential consequences of ocean acidification are not fully understood, but there are concerns that structures made of calcium carbonate may become vulnerable to dissolution, affecting corals and the ability of shellfish to form shells.

Specific Examples

- Chinese and Russian industrial pollution such as phenols and heavy metals in the Amur River have devastated fish stocks and damaged its estuary soil.
- Wabamun Lake in Alberta, Canada, once the best whitefish lake in the area, now has unacceptable levels of heavy metals in its sediment and fish.
- Acute and chronic pollution events have been shown to impact southern California kelp forests, though the intensity of the impact seems to depend on both the nature of the contaminants and duration of exposure.
- Ballast water taken up at sea and released in port is a major source of unwanted exotic marine life. The invasive freshwater zebra mussels, native to the Black, Caspian and Azov seas, were probably transported to the Great Lakes via ballast water from a transoceanic vessel.
- Due to their high position in the food chain and the subsequent accumulation of heavy metals from their diet, mercury levels can be high in larger species such as bluefin and albacore. As a result, in March 2004 the United States FDA issued guidelines recommending that pregnant women, nursing mothers and children limit their intake of tuna and other types of predatory fish.
- Some shellfish and crabs can survive polluted environments, accumulating heavy metals or toxins in their tissues. For example, mitten crabs have a remarkable ability to survive in highly modified aquatic habitats, including polluted waters. The farming and harvesting of such species needs careful management, if they are to be used as a food.

- Mining has a poor environmental track record. For example, according to the United States Environmental Protection Agency, mining has contaminated portions of the headwaters of over 40% of watersheds in the western continental US. Much of this pollution finishes up in the sea.
- Heavy metals enter the environment through oil spills - such as the Prestige oil spill on the Galician coast - or from other natural or anthropogenic sources.

Solutions

Marine pollution is part of the problem of too much pollution by humans in general. There are only two ways to remedy this: either the human population is reduced, or the ecological footprint left behind by the average human is reduced. If we do not follow the second way, then the first way may be imposed upon us, as world ecosystems falter and cease to support us.

The second way is for us, individually, to consume and pollute less than we do currently. For this there must be social and political will, together with a shift in awareness, so more people respect their environment and are less disposed to abuse it.

At an operational level, regulations, and international government participation is needed. It is often very difficult to regulate marine pollution because pollution spreads over international barriers, thus making regulations hard to create as well as enforce.

Perhaps the most important strategy for reducing marine pollution is education. Most are unaware of the sources, and harmful effects of marine pollution, and therefore little is done to address the situation. In order to inform the population of all the facts, in depth research must be done to provide the full scale of the situation. Then this information must be made public.

Oil Spill

An oil spill is the release of a liquid petroleum hydrocarbon into the environment due to human activity, and is a form of pollution. The term often refers to marine oil spills, where oil is released into the ocean or coastal waters. The oil may be a variety of materials, including crude oil, refined petroleum products (such as gasoline or diesel fuel) or by-products, ships' bunkers, oily refuse or oil mixed in waste. Spills take months or even years to clean up. Oil is also released into the environment from natural geologic seeps on the sea floor. Most human-made oil pollution comes from land-based activity, but public attention and regulation has tended to focus most sharply on seagoing oil tankers.

Environmental Effects

The oil penetrates and opens up the structure of the plumage of birds, reducing its insulating ability, and so making the birds more vulnerable to temperature fluctuations and much less buoyant in the water. It also impairs birds' flight abilities, making it difficult or impossible to forage and escape from predators. As they attempt to preen, birds typically ingest oil that coats their feathers, causing kidney damage, altered liver function, and digestive tract irritation. This and the limited foraging ability quickly causes dehydration and metabolic imbalances. Hormonal balance alteration including changes in luteinizing protein can also result in some birds exposed to petroleum. Most birds affected by an oil spill die unless there is human intervention. Marine mammals exposed to oil spills are affected in similar ways as seabirds. Oil coats the fur of Sea otters and seals, reducing its insulation abilities and leading to body temperature fluctuations and hypothermia. Ingestion of the oil causes dehydration and impaired digestion.

Because oil floats on top of water, less light penetrates into the water, limiting the photosynthesis of marine plants and phytoplankton. This, as well as decreasing the fauna populations, affects the food chain in the ecosystem.

Methods of Cleaning

A sheen is usually dispersed (but not cleaned up) with detergents which makes oil settle to the bottom. Oils that are denser than water, such as Polychlorinated biphenyls (PCBs), can be more difficult to clean as they make the seabed toxic.

Methods for cleaning up include:

- Bioremediation: use of microorganisms or biological agents to break down or remove oil.
- Bioremediation Accelerator: Oleophilic, hydrophobic chemical, containing no bacteria, which chemically and physically bonds to both soluble and insoluble hydrocarbons. The bioremedation accelerator acts as a herding agent in water and on the surface, floating molecules to the surface of the water, including solubles such as phenols and BTEX, forming gel-like agglomerations. Non-detectable levels of hydrocarbons can be obtained in produced water and manageable water columns. By overspraying sheen with bioremediation accelerator, sheen is eliminated within minutes. Whether applied on land or on water, the nutrient-rich emulsion, creates a bloom of local, indigenous, pre-existing, hydrocarbon-consuming bacteria. Those specific bacteria break down the hydrocarbons into water and carbon dioxide, with EPA tests showing 98% of alkanes biodegraded in 28 days; and aromatics being biodegraded 200 times faster than in nature.
- Controlled burning can effectively reduce the amount of oil in water, if done properly. But it can only be done in low wind, and can cause air pollution.

- Dispersants act as detergents, clustering around oil globules and allowing them to be carried away in the water. This improves the surface aesthetically, and mobilizes the oil. Smaller oil droplets, scattered by currents, may cause less harm and may degrade more easily. But the dispersed oil droplets infiltrate into deeper water and can lethally contaminate coral. Recent research indicates that some dispersants are toxic to corals.
- Watch and wait: in some cases, nautural attentuation of oil may be most appropriate, due to the invasive nature of facilitated methods of remediation, particularly in ecologically sensitive areas.
- Dredging: for oils dispersed with detergents and other oils denser than water.
- Skimming: requires calm waters
- Solidifying

Equipment used includes:

- Booms: large floating barriers that round up oil and lift the oil off the water
- Skimmers: skim the oil
- Sorbents: large absorbents that absorb oil
- Chemical and biological agents: helps to break down the oil
- Vacuums: remove oil from beaches and water surface
- Shovels and other road equipments: typically used to clean up oil on beaches

Prevention

- Secondary containment - methods to prevent releases of oil or hydrocarbons into environment.
- Oil Spill Prevention Containment and Counter measures (SPCC) program by the United States Environmental Protection Agency.
- Double hulling - build double hulls into vessels, which reduces the risk and severity of a spill in case of a collision or grounding. Existing single-hull vessels can also be rebuilt to have a double hull.

Environmental Sensitivity Index (ESI) Mapping

NOAA's Office of Response and Restoration

Environmental Sensitivity Index (ESI) maps are used to identify sensitive shoreline resources prior to an oil spill event in order to set priorities for protection and plan cleanup strategies. By planning spill response ahead of time, the impact on the environment can be minimized or prevented. Environmental sensitivity index maps are basically comprised of information within the following three categories: shoreline type, and biological and human-use resources.

Shoreline Type

Shoreline type is classified by rank depending on how easy the oil would be to cleanup, how long the oil would persist, and how sensitive the shoreline is. The floating oil slicks put the shoreline at particular risk when they eventually come ashore, covering the substrate with oil. The differing substrates between shoreline types vary in their response to oiling, and influence the type of cleanup that will be required to effectively decontaminate the shoreline. In 1995, the National Oceanic and Atmospheric Administration extended ESI maps to lakes, rivers, and estuary shoreline types. The exposure the shoreline has to wave energy and tides, substrate type, and slope of the shoreline are also taken into account – in addition to biological productivity and sensitivity. The productivity of the shoreline habitat is also taken into account when determining ESI ranking. Mangroves and marshes tend to have higher ESI rankings due to the potentially long-lasting and damaging effects of both the oil contamination and cleanup actions. Impermeable and exposed surfaces with high wave action are ranked lower due to the reflecting waves keeping oil from coming onshore, and the speed at which natural processes will remove the oil.

Biological Resources

Habitats of plants and animals that may be at risk from oil spills are referred to as "elements" and are divided by functional group. Further classification divides each element into species groups with similar life histories and behaviors relative to their vulnerability to oil spills. There are eight element groups: Birds, Reptiles and Amphibians, Fish, Invertebrates, Habitats and Plants, Wetlands, and Marine Mammals and Terrestrial Mammals. Element groups are further divided into sub-groups, for example, the 'marine mammals' element group is divided into dolphins, manatees, pinnipeds (seals, sea lions and walruses), polar bears, sea otters and whales. Issues taken into consideration when ranking biological resources include the observance of a large number of individuals in a small area, whether special life stages occur ashore (nesting or molting), and whether there are species present that are threatened, endangered or rare.

Human-Use Resources

Human-use resources are divided into four major classifications; archaeological importance or cultural resource site, high-use recreational areas or shoreline access points, important protected management areas, or resource origins. Some examples include airports, diving sites, popular beach sites, marinas, natural reserves or marine sanctuaries.

Estimating the Volume of a Spill

By observing the thickness of the film of oil and its appearance on the surface of the water, it is possible to estimate the quantity of oil spilled. If the surface area of the spill is also known, the total volume of the oil can be calculated.

Oil spill model systems are used by industry and government to assist in planning and emergency decision making. Of critical importance for the skill of the oil spill model prediction is the adequate description of the wind and current fields. There is a worldwide oil spill modeling (WOSM) program.

CPCB Water Quality Criteria for Surface Water Use

S.No.	Designated-Best-Use	Class of water	Criteria
1.	Drinking Water source without conventional treatment but after disinfection	A	* Total coliforms Organism MPN/100 ml shall be 50 or less * pH between 6.5 and 8.5 * Dissolved Oxygen 6mg/l or more * Biochemical Oxygen Demand 5 days 20°C 2mg/l or less
2.	Outdoor bathing (Organised)	B	* Total Coliform Organism MPN/100 ml shall be 500 or less pH between 6.5 and 8.5. Dissolved Oxygen 5mg/l or more. * Biochemical Oxygen Demand 5 days 20°C 3mg/l or less
3.	Drinking water source	C	* Total Coliform Organism MPN/100 ml shall be 5000 or less pH between 6 to 9. Dissolved Oxygen 4 mg/l or more. * Biochemical Oxygen Demand 5 days 20°C 3mg/l or less
4.	Propagation of Wild	D	* pH between 6.5 to 8.5. Dissolved Oxygen 4 mg/l or more * Free Ammonia (as N) 1.2 mg/l or less
5.	Irrigation, Industrial Cooling, Controlled Waste disposal	E	* pH between 6.0 to 8.5 * Electrical Conductivity at 25°C micro mhos/cm Max. 2250 * Boron Max 2 mg/l
		Below-E	Now meeting A, B, C, D & E Criteria

Index

E

F

G

H

I

K

L

M

N

O